Genetics Society and Clinical Practice

To our families

Genetics Society and Clinical Practice

Peter S. Harper and Angus J. Clarke
Institute of Medical Genetics,
University of Wales College of Medicine,
Cardiff, UK

© BIOS Scientific Publishers, 1997

First published 1997

All rights reserved. No part of this book may be reproduced or transmitted, in any form or by any means, without permission.

A CIP catalogue record for this book is available from the British Library.

ISBN 1 859962 06 8

BIOS Scientific Publishers Ltd
9 Newtec Place, Magdalen Road, Oxford OX4 1RE, UK.
Tel. +44 (0) 1865 726286. Fax +44 (0) 1865 246823
World-Wide Web home page: http://www.Bookshop.co.uk/BIOS/

DISTRIBUTORS

Australia and New Zealand
Blackwell Science Asia
54 University Street
Carlton, South Victoria 3053

India
Viva Books Private Limited
4325/3 Ansari Road, Daryaganj
New Delhi 110002

Singapore and South East Asia
Toppan Company (S) PTE Ltd
38 Liu Fang Road, Jurong
Singapore 2262

USA and Canada
BIOS Scientific Publishers
PO Box 605, Herndon
VA 20172-0605

Production Editor: Rachel Robinson.
Typeset by Poole Typesetting (Wessex) Ltd, Bournemouth, UK.
Printed by Biddles Ltd, Guildford, UK.

CONTENTS

ABOUT THE AUTHORS

Many books include a few sentences about the authors of a book, often on the back cover, where it may be difficult to decide whether they form part of the book itself or of the associated advertising. Such notes are usually brief, polite and tell one little about the authors' wider characteristics.

Since this book has been so largely shaped by our own personal views – some would say prejudices – we feel we owe it to readers to say a little more about ourselves than is customary. We hope that some information about our background may help others understand why we write and feel as we do. Whether these short pen-portraits, written by ourselves, in any way reflect our true characters, we leave to those who know us well to decide.

Peter Harper

Professional. Peter Harper studied Medicine and Biology at Oxford University, where he developed his interest in basic genetics, with subsequent clinical training in London. After qualifying in Medicine in 1964, he held a variety of posts in internal medicine, paediatrics and neurology, before moving to Liverpool to work with Professor Cyril Clarke at the Nuffield Institute of Medical Genetics. This was followed by a 2 year Fellowship at Johns Hopkins Hospital, Baltimore, under Dr Victor McKusick.

On returning to the UK, he set up a new Medical Genetics unit in Cardiff, where he is now Professor of Medical Genetics as well as Consultant Physician and Clinical Geneticist. Inherited neurological disorders have always been his main field, first from a clinical genetics perspective and later involved with molecular genetics. Much of his interest in social and ethical issues has arisen from his practical experience with genetic counselling and predictive testing for Huntington's disease families. His other books include *Practical Genetic Counselling* (4th Edn, 1993), *Huntington's Disease* (2nd Edn, 1996), *Myotonic Dystrophy* (2nd Edn, 1989) and *Population and Genetic Studies in Wales* (1986).

Personal. Born in 1939 into a strongly medical family in a small country town (Barnstaple, Devon), he developed an early fascination for natural history, with a broader interest in evolution stimulated by his father, who was also a physician. Living in Cardiff, Wales, for the past 25 years, he is married with five children (three adopted). He has no religious affiliations and has never belonged to a political party, but is a determined defender of the UK National Health Service and would regard himself as firmly on the 'left' politically.

Angus Clarke

Professional. Angus Clarke studied Medical and Natural Sciences at King's College Cambridge, taking the part II course in Genetics in his final year. He then studied clinical medicine at Oxford for 3 years, qualifying in 1979, and worked in junior hospital posts for 2 years. He trained for 4 years in paediatrics, and then – seeking experience in research – he took a position in Medical Genetics with Peter Harper. This entailed visiting families from all over Britain with a rare condition called X-linked hypohidrotic ectodermal dysplasia. Many of these families reported ways in which social intolerance had added to their burdens, and even health professionals had sometimes added to their problems instead of relieving them. Getting to know some of these families as personal friends has certainly influenced the direction of his work ever since. After 2 years' experience in Newcastle, working in both clinical genetics and paediatric neurology, he returned to Cardiff as a Senior Lecturer, now Reader, in Clinical Genetics.

In drawing attention to the human problems potentially caused by genetic counselling or genetic screening programmes, Angus Clarke is anxious that medical genetics does not promise more than it can deliver – that it knows its limitations. His views have arisen from reflecting on the accounts of the experiences that have been described by families he has met, and are not determined by any particular theoretical position.

Personal. Angus Clarke comes from a medical family – both his parents are doctors – and was brought up in north London with a sense that he belonged elsewhere because of family roots in Scotland and south Wales. In summer holidays from school and university he travelled widely in Europe, the Moslem world and Asia; he also spent 3 months in Sudan during his clinical studies. He has been a member of at least two political parties that he can recall (Labour and the Green Party), but is not politically active now except from his armchair. He has a strong commitment to the National Health Service. He was brought up in an Anglican household, but had no active religious involvement for many years. He has been a Quaker and a vegetarian for about 10 years, both these positions arising out of his views rather than accounting for them.

ACKNOWLEDGEMENTS

Many of the chapters in this book are based on papers that have been previously published, or are in press, in a variety of journals. We should like to thank the editors and publishers involved for their permission to reproduce material, in particular the *Lancet*, *The British Medical Journal* and the *Journal of Medical Genetics*. Specific acknowledgements of sources are given in the individual chapters.

We owe a particular debt to our colleagues in Cardiff for sharing their ideas and thoughts, many of which are incorporated here, though we ourselves are responsible for any views expressed. In particular, we thank members of our Genetics and Social Sciences Research Group, including Julie Binedell, Kate Brain, Lucy Brindle, Jonathon Gray and Evelyn Parsons, as well as our consultant colleagues Sally Davies, Helen Hughes, David Ravine and Julian Sampson. Professor David Cooper deserves a special thanks for having linked us with BIOS, to whose editorial staff we are grateful for their speedy production and tolerance in allowing late changes.

GENERAL INTRODUCTION

Why do we feel that a book on the topic of 'genetics, society and clinical practice' is needed? There is no shortage of articles on a wide variety of social and ethical aspects of human genetics, and indeed much of the material on which our book is based has previously been published in medical or genetic journals. Several valuable books already exist, some orientated towards professionals, others for a broader public consumption. Yet when we began to plan this book 3 years ago, we were conscious of a need, and of a gap in the current literature; we feel that the need has grown since that time. What is this need, and why is there a gap?

This question is best answered by attempting to explain what the present book aims to be and to do, and why we have written it ourselves. We have tried to cover important ethical and social issues arising in *clinical genetics practice*. This practice, with which both of us have been involved on a day-to-day basis throughout our professional careers, involves genetic counselling and diagnosis; the application of genetic tests in diagnosis, in carrier detection and in pregnancy; the co-ordination of screening programmes, and the teaching and explanation of all these fields to students, to professional colleagues and to the broader public.

As a result of our work in these areas, partly research, partly service, we have encountered all of the issues that form the basis for the different chapters in this book, and have seen at first hand the problems that can arise for families with genetic disorders if clinicians and others involved are unaware of them or handle them insensitively. Most of the papers forming the core of the book were written because we felt that many people were *not* aware of the importance of the issues; we wished to place the particular topic 'on the agenda', so that professional and wider debate might allow some general consensus to emerge as to what could be considered 'good practice' and so that professional guidelines could be drawn up. For some of these issues, such as the genetic testing of children for late-onset disorders, it is probably fair to say that an agreed code of practice has, after considerable debate, emerged; most professionals in genetics, at least in Europe and America, are aware of the potential problems, and other clinical specialties are, more slowly, also recognizing them. In re-reading our original papers, we have been struck by how little those selected for inclusion here have dated; probably this is because they have dealt principally with general issues, which remain as relevant today as when the papers were first written.

This book thus concentrates on the practical issues that affect individuals in real life, but we are conscious that there are other areas, often attracting much more media attention, that we have not tried to cover at all. These include commercial topics such as gene patenting, ethical dilemmas arising from gene therapy (germ-line or otherwise), or the transfer of human genes to other species and the possible cloning of humans. Our excuse for leaving aside these and other topics is that, at present, they are of little relevance to the immediate problems faced by families affected by genetic disorders; likewise we have omitted some other areas, such as the field of pre-implantation diagnosis, because we have no special experience in this area.

We are very conscious that in writing a book which deals with genetics and society, we cannot claim to be social scientists of any form – we are not formally trained in psychology, sociology, philosophy or ethics. Experts in these academic disciplines may think (indeed some have told us!) that this should disqualify us at once from presuming to write a book impinging on these fields, or from attempting to undertake any work in them. Our answer would be that we work closely with colleagues skilled in all these fields, that their involvement is essential in any detailed analysis, but that in recognizing and creating wider awareness of the issues we are well placed because of the strongly practical base of our professional work. We might be tempted to add that most social and behavioural scientists have until recently been as ignorant about clinical genetics and its problems as we originally were about the social sciences! We hope that our own papers and this book can help to encourage the growing links between the disciplines which are beginning to prove so fruitful.

We should particularly like to emphasize that this book is a reflection of our personal views and experience and does not attempt a complete review of the overall literature on the particular topics. It should be regarded more as a collection of essays on a theme than as a textbook. Despite this, we think that our experience, and frequently our views, will not be far removed from those of many others involved with families affected by genetic disorders. Likewise, while our practice is based in the UK, we think that the issues are likely to be of general relevance to those working in other countries, particularly other parts of Europe; probably also North America and the 'Western world' generally. We recognize that social structures and attitudes to genetic services will differ greatly in many developing countries, but the themes set out here are equally relevant in our view to all countries, even though the approaches will necessarily differ.

We have grouped the topics of this book in four main sections, each of which is introduced separately. Section I covers some of the major issues involving genetic testing, the most rapidly growing area of medical genetics and perhaps the one containing the most obvious and immediate problems. Topics such as the genetic testing of children and the growing field of testing for late-onset genetic disorders are practical and important issues affecting clinical practice, while the controversial areas of 'over the counter' genetic tests and insurance and genetics raise questions as to the need for regulation and legislation.

Section II, on population screening for genetic disorders, links closely with the preceding section on genetic testing and also relates to the discussion on genetics and public health in the subsequent section, raising the need for awareness of the different and potentially conflicting goals of those whose work relates mainly to individuals and families (including the authors), and those who see genetic developments primarily in terms of their population and public health value.

The third section raises some controversial topics in which we have found ourselves involved and not always in tune with some of our clinical and research colleagues. We strongly believe, however, that a questioning and critical attitude, both to new developments and to established concepts, is important and necessary in a field like medical genetics, which inevitably impinges on so many controversial areas. We feel that it is much healthier for this questioning to come from within the genetics community, rather than for those in the field to 'close ranks' against external criticism. There is all the more reason to maintain a critical attitude in the light of the disastrous abuses that have already been carried out in the name of genetics by professionals and by entire social systems in the past. This theme of the abuse of human genetics forms the subject of the final section of this book. If genetics is to play a beneficial role in our present society, we need to keep alive our awareness of these past abuses and to maintain our vigilance that new developments are not misused in the future.

Section I

ISSUES IN GENETIC TESTING

This section covers some of the general issues relevant to genetic testing. The opening chapter asks the question 'What do we mean by genetic testing?', aiming to resolve some of the confusion that surrounds this, so that genuine issues rather than terminology can be debated. It also sets out a theme which recurs throughout the book – the importance of different genetic services and processes being closely integrated, rather than seen as compartmentalized activities. This is nowhere more clearly illustrated than in genetic testing, with the need for close links between clinical, laboratory and psychosocial aspects.

Chapter 2 explores the specific topic of genetic testing of children for 'adult' disorders, one whose implications were largely ignored until our early *Lancet* paper of 1990 put this on the agenda. Since then, there has been considerable debate, but also a tendency at times for attitudes to be polarized; we have always felt this to be unfortunate and hope that the present chapter makes our support for a flexible and individual approach clear.

The third chapter concentrates on some of the special problems arising in the context of genetic prediction for late-onset disorders, using Huntington's disease as a starting point. It is fortunate in a number of ways that our experience of this disorder has come first and has been so thoroughly documented, as it poses in clear form most of the difficulties and issues involved with prediction and presymptomatic testing. Increasingly, those involved with prediction in other disorders have realized that they need to face up to the same situations, and that Huntington's disease has valuable lessons for the whole field of late-onset genetic disorders.

The fourth chapter in this part tackles the controversial issue of insurance in relation to genetic testing. This is examined very much from a UK perspective in which life insurance rather than health insurance is at present the predominant factor. Perhaps the main general point that arises is the widespread concern over disclosure of sensitive genetic test information and the fear that once such information is given to a third party or becomes part of a formal medical record it could be passed on to others. This fear links with the final chapter in the section, dealing with the relatively new concept of 'over the counter' testing, where we could be on the threshold of an approach which entirely excludes professionals and whose effects, for that very reason, would be almost impossible to evaluate. At present, the true demand for this would seem to be very small; whether it increases may depend directly on whether people are forced to disclose genetic test results and whether they are able to trust that their medical records are truly confidential.

Chapter

1

WHAT DO WE MEAN BY GENETIC TESTING?

P.S. Harper

INTRODUCTION

Genetic testing is a topic that stimulates widespread discussion and debate, not just among genetics professionals, but among clinicians and scientists generally, and increasingly involving the wider public. Views are expressed in the scientific and general press, and through other media, about the likely benefits and dangers that may result. In the UK there is now an official government body, the Advisory Commission on Genetic Testing [1], with a counterpart in the United States, the Task Force on Genetic Testing, which has recently produced a report and recommendations [2]. There has been a *Journal of Genetic Testing* recently launched.

As someone involved in the practice of genetic testing as a clinical geneticist, and also in some aspects of the general debate on the topic, I have frequently found that the term 'genetic testing' means different things to different people, and that this can confuse any discussion on its practical consequences and general implications. Since part of this confusion has at times existed within my own mind, I have been forced to think through the issues involved and hopefully to clarify them to some extent. Perhaps others will find it helpful to read some of my conclusions, even though they are largely personal views.

SOME DEFINITIONS

There are several possible definitions of the term genetic testing, each of which may be appropriate in one particular context, but unhelpful in others. They largely depend on the meaning attached to the word 'genetic'. It should be noted that the term 'genetic screening' is avoided entirely here, since 'screening' implies application to large population groups, which is the exception in

---

This chapter is based on a 'commentary' in the *Journal of Medical Genetics* (1997), Vol. 39, and appears here by kind permission of the BMJ Publishing Group.

relation to most current genetic testing activities, though this will change in the future. The special issues associated with screening are considered in Section II of this book.

The first definition of genetic testing is based on a technological interpretation, and would include any test involved in analysis of the genetic material, whether involving somatic cells or the germ-line. Such a definition could include the numerous changes seen in the genes and chromosomes of cancer cells, mostly non-heritable, as well as DNA analysis of pathogenic organisms involved in human disease, which plays an increasing role in medical diagnostics. Such tests, however important, are not usefully considered together with tests involving potentially heritable genetic changes; they do not raise the same social and ethical issues, and they can properly be left along with other laboratory medical tests, to be assessed, costed and implemented (or not) on their merits.

If a more restricted definition of genetic testing is accepted, confined to the germ-line and specifically to the human germ-line, the question arises – why is a test any less genetic because it may be analysing the product or function of the gene rather than the gene itself? There are good reasons for not restricting our definition to the analysis of DNA, since testing procedures may shift between different technologies without altering the fundamental issues involved. Thus, for fragile X mental retardation, testing which originally involved microscopic chromosome analysis is now based on detection of the specific DNA mutation, but may well in future utilize the protein product of the gene [3]. All three approaches can reasonably be considered as genetic testing, and the relevant factor is not the technology, but the fact that the test is detecting a change directly related to an inherited disorder.

For purposes of discussion relating to human inherited disorders and the surrounding medical social and ethical issues, a useful working definition might be along the following lines:

> Genetic testing is the analysis of a specific gene, its product or function, or other DNA and chromosome analysis, to detect or exclude an alteration likely to be associated with a genetic disorder.

Perhaps the most important consideration is to make sure that the definition used is one appropriate to the activities and issues that one wishes to cover.

WHY IS GENETIC TESTING DIFFERENT FROM OTHER FORMS OF MEDICAL TEST?

This is a question that is often asked; indeed insurance bodies have specifically denied that it is different.

For "known genetic diseases such as Huntington's chorea, cystic fibrosis or Duchenne muscular dystrophy . . . the insurance industry already has experience of individuals who have had a genetic test because of medical history and insurers treat the report of such tests in exactly the same way as the results of any other medical test." [4]

Such a categorical statement makes it wise to ask whether genetic testing is actually a medical investigation at all? In some circumstances it is clearly not. DNA 'fingerprinting' for paternity testing or forensic analysis provides an example, while study of the normal variations in human proteins and DNA used in anthropological and population genetics is likewise not regarded as a medical activity, even though such studies may involve human genes that may be as significant as those known to underlie specific genetic disorders. That this is a line that can easily be crossed is seen by the issues raised in testing for genetic variants of apolipoprotein E (ApoE), widely used as a population marker, but now shown to have important correlations with susceptibility to Alzheimer's disease [5] and cardiovascular disease [6]. It is possible, indeed likely, that other markers previously considered 'neutral' may prove to have important disease associations.

At the other extreme, many diagnostic genetic tests can readily be accepted as 'medical' in nature, We have, however, to be clear as to what we mean by 'diagnostic'. I would restrict the term to those genetic tests done on individuals who either have, or are suspected of having a particular disorder because of clinical symptoms or signs, and where the question to be answered is whether the patient has the particular disorder *at present*, not whether they may develop it at some time in the future. It is increasingly possible for genetic testing to answer this question in the absence of any known family history. Such tests will now often be requested by clinicians as part of their regular clinical practice, not just by those trained in clinical genetics. These tests will also increasingly replace other less accurate or more invasive (and often more costly) medical investigations.

It is extremely important to ensure that 'diagnostic genetic testing', as outlined above, is not confused with 'predictive' or 'presymptomatic' genetic testing, discussed below, where the individuals concerned are usually not 'patients' at all, but are healthy individuals. Even where symptoms are present, they must be relevant to the disease if testing is to be considered diagnostic. Thus, a test for the Huntington's disease (HD) mutation undertaken by a neurologist because of a movement disorder, could be considered as 'diagnostic', whereas the same test carried out on a patient with headache (not a feature of the disorder) and a family history of HD, would not be 'diagnostic' but 'predictive'. Serious consequences will result if these fundamental differences in applying identical technology are not understood.

It is in the category of 'presymptomatic' and 'predictive' testing that most of the difficult issues involving genetic testing lie. Again, clear definitions are important here; 'presymptomatic' testing is best reserved for those situations where an abnormal test result will almost inevitably lead to development of the disease at some point in later life (e.g. HD), whereas the term 'predictive testing' covers a broader range of situations in which the risk of a disorder occurring is substantially increased or reduced, but without necessarily implying any degree of certainty.

Whether these forms of genetic testing are or should be considered 'medical' in nature, and regulated as such, raises fundamental issues that have not been fully debated, let alone resolved. It may well be that there are strong reasons for medical involvement in some situations, but I see no reason why it should be assumed that all genetic testing in this group should automatically be 'medicalized', whether by insurance companies, the medical profession or others. One could perhaps point to the example of normal pregnancy as a process which has become medicalized without the full balance of benefit and harm being adequately assessed. Already one can see a progressive medicalization in the language that is often used in relation to genetic tests. Thus, healthy individuals undergoing predictive testing are often referred to as 'patients', while those with an abnormal test result are called 'affected', even though onset of the disorder may be many years away, or may not occur at all [7].

THE GENETIC TESTING PROCESS

Whatever we think of the matter, genetic testing is here to stay, both in diagnosis and prediction. It has been part of clinical genetics practice for almost a decade, and is now becoming widespread in the practice of most specialties. How can we ensure that it is delivered as efficiently and appropriately as possible, and to the highest standards? Perhaps the most important factor in this will be to ensure that genetic testing is looked at as an overall process, not simply as a laboratory activity.

Many lay people's concept of a test is of something that starts when a sample is taken and ends when a result is produced, but this is far from the case and in no situation is it more important to recognize this than in the process of genetic testing. This process can be illustrated diagramatically in *Figure 1.1a*, where the laboratory aspects of the test are represented in the central portion (B), while the equally important aspects of preparation, information and consent, and the subsequent aspects of interpretation and support are shown as the outer portions (A and C).

Naturally the central laboratory portion of the testing process is crucial to the whole and needs to be assessed in terms of scientific validity, accuracy, efficiency and cost, all of which form part of what is usually considered 'quality control'. As genetic testing moves from being a research activity to an established service, these issues are being addressed, often on a national or international basis. Improved technology, cost pressures and increasing commercial involvement are all powerful factors ensuring progress. The danger, however, is that the evolution of this part of the testing process may become detached from the essential preceding and subsequent parts, unless we think about, plan, evaluate and cost the testing process as a whole.

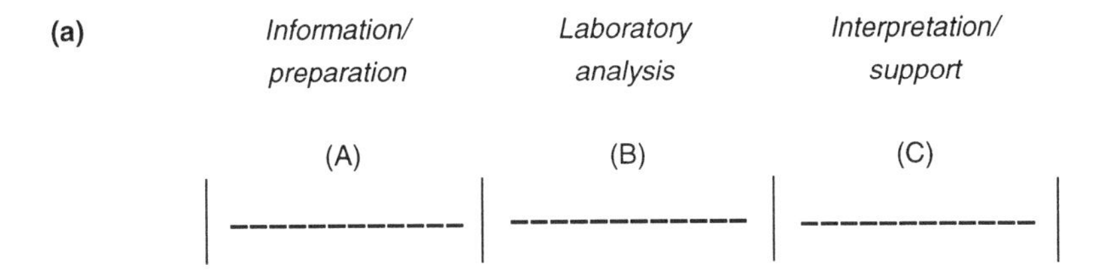

(b)

1. Non-laboratory aspects relatively simple or restricted (e.g. cystic fibrosis carrier screening)

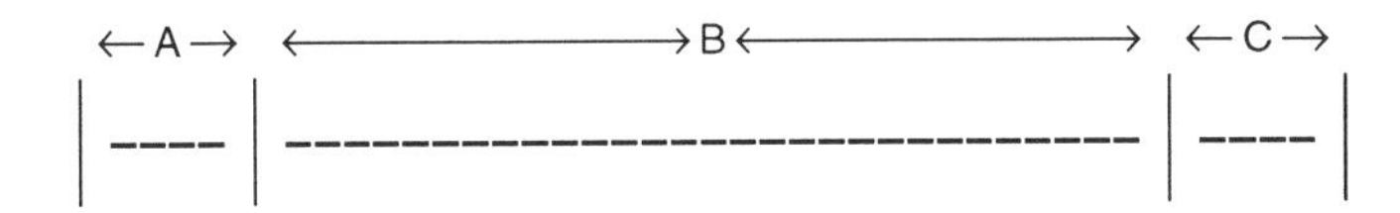

2. Non-laboratory aspects complex (e.g. Huntington's disease presymptomatic testing)

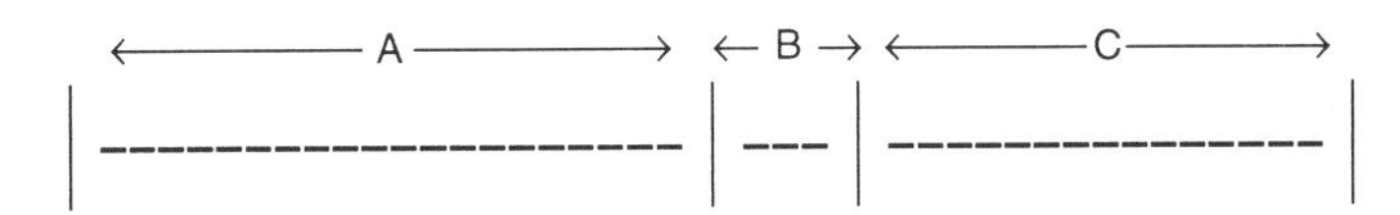

Figure 1.1. The genetic testing process. (a) Essential components of genetic testing; (b) variation of components in different testing situations.

In proposing an integrated approach to genetic testing, I would argue that there is ample precedent for this and that it is dangerous, even negligent to do otherwise. Precedent comes from several sources: thus in testing for human immunodeficiency virus (HIV) it is generally accepted that laboratory testing should only be done in the context of counselling, while much of the skill of those working in laboratory medicine as pathologists lies in the interpretation of results in a clinical context and the determination of which investigations are most appropriate; pathologists have rightly warned against the dangers of 'results only' test procedures [8].

If we think of genetic testing as analogous to a surgical procedure, one could reasonably ask what body proposing or commissioning such an operative procedure would do so unless the framework of preoperative assessment and post-operative care were to have been evaluated, costed and put in place to the same standards as the operation itself. Any surgeon or hospital failing to do this would rightly be considered negligent, as would any hospital that, on grounds of cost, decided to provide only the operation, not the pre- or post-operative care.

Of course it is unnecessary and unrealistic for all genetic testing procedures to be surrounded by a forest of associated activities. As indicated in *Figure 1.1b*, there will be many situations, especially in diagnostic or carrier testing, where these aspects will be quite limited, and where written information may appropriately replace expensive professional time. In other genetic testing situations, however, it may prove that the associated risk estimation, preparation, information and support, all part of what is generally recognized as 'genetic counselling', will be as important and as or more expensive, than the actual laboratory procedure. Presymptomatic testing for late-onset disorders, such as HD and familial cancers, are cases in point. The balance will clearly vary between different situations as will the question of whether specialists in clinical genetics, other hospital clinicians, or those in primary care should be responsible; what is essential is that all the aspects are considered together as integral parts of the genetic testing process.

THE FUTURE OF GENETIC TESTING

Those who have expressed concerns about genetic testing have been worried, not so much about the present, but about the future. These concerns have included such issues as: what will happen when genetic testing becomes widespread and commonplace? Will commercial interests, new technology and cost pressures on clinical time result in numerous tests being done without the understanding and consent of those involved? How will access to genetic test results be controlled? Will testing result in widespread discrimination in employment, insurance and access to health care? All these are valid concerns and it is as well to address them now, before genetic testing becomes more widespread.

Some with whom I have discussed these aspects are fatalistic and see widespread misuse as inevitable. I disagree, and take the view that – providing professionals in the field, both clinicians and scientists, take the lead in setting standards of practice, and involve both the general public and government bodies in this – we should be able to avoid the widespread abuse of genetic tests that might otherwise occur. I can point to some hopeful signs.

A small but significant pointer is our experience of presymptomatic testing for HD, an activity that was widely forecast as likely to have disastrous effects. Over 2000 such tests have now been carried out in the UK alone, [9] and the experience of most countries, including the UK, Canada, The Netherlands and Belgium is that harmful effects have been few [10,11] and that most people tested, regardless of the result, have felt a benefit from being tested. Widely accepted international guidelines have been produced and followed (see Chapter 3), and it has generally been accepted that presymptomatic testing in other serious late-onset disorders should follow comparable lines. Given the gravity of the issues involved and the potential for disaster, I regard this result as a considerable success. The challenge is how to maintain standards as testing encompasses increasing numbers of diseases and as pressures to contain costs of genetic services increase, particularly since the cost of pre- and post-test activities may equal or even exceed the laboratory costs of the test itself.

A second favourable pointer is the establishment of regulatory bodies, such as the Advisory Committee on Genetic Testing in the UK and the US Task Force on Genetic Testing. Such bodies may lack legal powers, but the influence of the guidelines they produce, and the likely consequences of ignoring them, are considerable. Such influences extend well beyond their countries of origin and can encourage good practice in those countries where there may be less awareness of the issues involved.

Can commercial pressures be overcome? Again I am optimistic that this can be achieved, with one important proviso – that commercial interests, whether pharmaceutical, technological, or insurance-related, are not in unfettered control of health or research policies. Provided that governments do not give up their regulatory or planning powers completely, I think it likely that most commercial organizations will generally prefer to work within a framework acceptable to governments and society. Although the less responsible may be influenced by the desire to achieve volume rather than appropriateness in genetic testing, the recognition that a few lawsuits could neutralize large profits is likely to be a powerful reminder of the advantages in maintaining responsible codes of practice.

I am in no doubt that, even 10 years from now, the patterns of genetic testing that will have evolved will be widely different from those that can be predicted now, and that what we mean by the process will equally have changed. By thinking and planning ahead, by trying to set high standards of practice now, and by insisting on rigorous evaluation of all aspects of genetic testing, we should be able to ensure that current concerns over its future misuse do not become reality.

REFERENCES

1. Government response to the third report of the House of Commons Select Committee on Science and Technology, 1994–95 session. January 1996.
2. Task Force of the NIH-DOE Working Group on ethical, legal and social implications of genome research (1997) *Promoting Safe and Effective Genetic Testing in the United States. Principles and Recommendations*. NIH-DOE, Baltimore, MD.
3. Willemsen R, Mohkaonsing S, De Vries B *et al.* (1995) Rapid antibody test for fragile X syndrome. *Lancet*, **345**: 1147–1148.
4. Association of British Insurers (1992) *Memorandum to Nuffield Council Working Party on Genetic Screening*. Association of British Insurers, London.
5. Saunders AM, Strittmatter WJ, Schmechel D *et al.* (1993) Association of apolipoprotein E allele e4 with late onset familial and sporadic Alzheimer's disease. *Neurology*, **43**: 1462–1472.

6. Davignon J, Gregg RE, Sing CF (1988) Apolipoprotein E polymorphism and atherosclerosis. *Arteriosclerosis*, **8**: 1.
7. Genetic testing for familial hypertrophic cardiomyopathy in newborn infants. An ethical debate. *Br. Med. J.* (1995) **310**: 856–858.
8. Royal College of Pathologists (1995) *Guidance for Purchasers of Pathology Services*. Royal College of Pathologists, London.
9. UK Huntington's Disease Prediction Consortium (1995). Presymptomatic testing for Huntington's disease in the UK. *Am. J. Hum. Genet.* **57**: A295.
10. Wiggins S, Whyte P, Huggins M *et al.* (1992) The psychological consequences of predictive testing for Huntington's disease. *New Engl. J. Med.* **327**: 1401–1405.
11. Decruyenaere M, Evers-Kiebooms G, Boogaerts A, Cassiman JJ, Cloostermans T, Demyttenaere K, Dom R, Fryns JP, van der Berghe H (1996) Prediction of psychological functioning one year after the predictive test for Huntington's disease and impact of the test result on reproductive decision making. *J. Med. Genet.* **33**: 737.

Chapter 2

THE GENETIC TESTING OF CHILDREN

A.J. Clarke

INTRODUCTION

The importance of genetic diseases in childhood has grown over the course of this century as that of infectious diseases and malnutrition has diminished, at least in developed countries. Accordingly, many of the children now requiring medical attention are recognized as suffering from genetic conditions. Making the diagnosis of a genetic disorder in an affected child is therefore a standard clinical activity. Establishing such a diagnosis has traditionally depended upon finding evidence of the disease process but is increasingly turning to genetic investigations that identify the underlying, genetic cause of the disorder. Such genetic tests can be carried out on any tissue and at any age, from conception onwards, because they do not depend upon the condition having manifested itself in any way. It is therefore possible to identify healthy children who will, in the future, develop a genetic disease; it is also possible to identify those who are healthy carriers of a disease-related gene which will never affect them but which may be of relevance to their own future children.

A sick child warrants full investigation to determine the cause of their illness, to ensure that the child is given the correct treatment and managed appropriately; this may entail genetic tests. This does not mean, however, that it is always helpful to identify healthy children likely to develop or transmit genetic disease in the future. Even diagnosing an affected child as having a genetic disease, as opposed to any other type of disease, may have serious repercussions which require careful consideration as a family becomes aware of them.

The fact that the disease affecting a child is found to be genetic may have far-reaching consequences for the family. The nature and extent of these consequences will often depend upon the mode of inheritance of the condition, which

---

This chapter is reproduced by kind permission of Kluwer Academic Publishers. It appears as the chapter 'Childhood testing: ethics and practice' in R. Chadwick (ed.) *The Ethics of Genetic Screening*. © 1998 Kluwer Academic Publishers, Amsterdam.

will determine which other family members will be at risk of developing the condition themselves or of transmitting it to their children. For example, if it is transmitted as an autosomal recessive trait, then the child's present or future siblings may develop the same disease but it would be less likely for the child's own future children to do so. Information like this will often be unwelcome but it is generated simply by establishing the diagnosis in the affected individual and so it cannot be avoided. Furthermore, members of the family will sometimes find the information to be helpful if it allows them to avoid complications of the disease or to take the risk of it into account when planning future children.

When a child presents with symptoms or signs of an illness that may be genetic in origin, therefore, any test carried out to establish the diagnosis may have important consequences both for the child and for other members of the family. This will be true whether or not the techniques used for establishing the diagnosis are genetic. It will always be important for clinicians to recognize these possible consequences and to help the child and the family adjust to their new situation, once the diagnosis has been made. Indeed, this is one of the principal tasks of clinical geneticists and genetic counsellors.

A very different situation arises when a family or a health professional considers the question of testing a healthy child to find out about his or her genetic constitution. This could arise in at least three different contexts:

(i) predictive testing to see if the child will develop a disease recognized in another family member (parent, uncle, aunt, cousin, brother or sister . . .);
(ii) carrier testing to see if the child may be at risk of having an affected child him- or herself when older;
(iii) a screening test unrelated to family history to identify children with increased susceptibility to some common disease such as ischaemic heart disease or diabetes.

The scope of all these types of testing has increased dramatically over the past 10 years, and will continue to increase as more genes are identified in the course of the Human Genome Project. These tests can be carried out on minute samples of any tissue from the body (even a mouthwash) and do not depend upon the child having developed the disease; they recognize the child's genotype and do not depend upon the phenotype (manifestations of the disease process). In this chapter, it is the issues raised by the genetic testing of healthy children that will be examined.

PREDICTIVE TESTING

When will it be helpful to carry out genetic testing to see if a healthy child will or will not develop a specific disease that runs in the family?

This type of predictive genetic testing will clearly be appropriate if the child is at risk of complications of the disease that can be avoided or ameliorated if recognized early. For example, if a parent has familial adenomatous polyposis coli (FAP) (an autosomal dominant condition predisposing to cancer of the large bowel) their child will be at a 1 in 2 (50%) risk of developing the same condition. Surveillance for tumours in FAP is often started at 10–12 years of age, so predictive testing before then would be appropriate to distinguish those children actually at risk of tumours, and in whom surveillance would be useful, from those who could be spared the unpleasant annual colonoscopy.

It may also be helpful to the family as a whole to carry out predictive testing if a child is at risk of developing a serious illness in childhood. There may be an affected brother or sister, or perhaps an uncle or cousin, and the family may be looking for early signs of the same condition in another child; they may be in emotional agony, watching the child to see if he develops early signs of the same condition. For example, if a boy has a progressive, sex-linked disorder such as Duchenne or Becker muscular dystrophy, then his younger brother may be watched intently to see if he shows any of the early signs of muscle weakness. It can be very helpful to test the younger boy to find out definitely whether he does or does not have the condition. If he is affected then he is no worse off; if he is unaffected then the family will be spared their frequent misinterpretations of fatigue or cramps as early signs of the same condition.

In other contexts, however, where the disease does not usually present until well into adult life and where there is no useful medical intervention to be made, the predictive testing of a child could have damaging consequences. Before predictive testing became possible for Huntington's disease (HD) (a disease causing memory loss, personality change and a movement disorder) there was debate about the circumstances in which such testing should be offered. Craufurd and Harris raised this as a concern [1], and a consensus was reached by the World Federation of Neurology and the International Huntington's Association (a federation of lay support groups) not to test children [2]. This was reinforced by the considered opinion of others involved in predictive testing for HD [3].

The grounds for such caution in relation to predictive testing of healthy children for late-onset disorders are essentially four-fold:

(i) if a child is tested then they lose the opportunity to make their own decisions in the future as an autonomous adult.

(ii) When an adult is tested, they have control over the result and how it is used and disseminated; even the fact of their being tested is treated with strict confidentiality. A child who is tested forfeits confidentiality in respect of both the fact of being tested and the test result.

(iii) The family's knowledge of the test result could result in the child being treated differently, and this might lead to social or emotional problems in the child. A child destined to develop the disease may be brought up with lower expectations for their education and career or altered expectations for their future relationships. Such expectations could then become self-fulfilling prophecies, and could damage the self-esteem of the child at the deepest levels. Even a favourable result may have difficult emotional consequences; a child shown not to carry the family's disease gene may feel excluded from the inner circle of family concerns. Problems within the family may be particularly likely if the test result conflicts with the result anticipated by family consensus in the context of 'pre-selection' [4].

Experience with predictive testing for HD in adults has shown that problems arise in a high proportion of cases. Most of these problems are not technical, laboratory problems but relate to counselling issues: the appropriateness of the test request, the attitudes of other family members or the disclosure of the test results [5]; Chapter 3 discusses these in greater depth. Furthermore, problems are as likely after a favourable as an unfavourable result, although the adverse consequences may be delayed for some months especially when the result has been favourable [6–9]. Particular problems could arise if more than one child in a sibship were tested, and they were given different results. This could easily lead to great difficulties in the relationships within the family, between the parents and children and among the children. The possibility of adverse consequences for insurance or employment in the future would also need to be considered, because an individual found to have the HD gene mutation might find themselves to be excluded from a career or from health or life insurance as if already affected by the disease, despite being perfectly healthy at that stage. Given that the problems experienced in the testing of adults have arisen despite intensive prior counselling of the tested individual, and that an individual to be tested in early childhood would often not be able to participate in such counselling, then there are good grounds for caution in this area.

Genetic counselling has, for good reasons, developed an adherence to the view that genetic testing should be linked to counselling of the individual to be tested in advance of the test itself; the experience with testing for HD has reinforced this tenet of the profession. Testing children would run counter to this principle, and the potential additional problems anticipated if children were to be tested combine to make such predictive testing most unwise.

(iv) An additional reason for commending caution towards predictive testing in such circumstances is that a majority of adults at-risk of HD choose not to be tested. Whereas a clear majority of at-risk family members stated hypothetically that they would seek testing once it became available, in fact only 10–15% have done so; when faced with the option in reality, most have chosen to retain uncertainty [10–13]. When a parent considers whether or not they wish their child to be tested for such a condition, they may respond as if they were confronted with the question hypothetically for themselves; the reality of the question from the perspective of the child may not have the same impact on the parent.

These considerations are not just abstract concerns, but the issues raised have had to be considered in many centres where requests for the predictive testing of children at risk of HD have been received from parents, adoption agencies and other social or legal bodies [5,14,15]. Furthermore, these issues do not arise just in relation to HD but also in relation to an increasing range of other disorders [16].

Given the small number of adults who in practice choose to be tested for HD when the test is available and then are confronted by the issues, and given that the rights of the child as a future adult to both autonomy and confidentiality are abrogated by testing in childhood, it is clear why so many professionals and lay groups [17] are opposed to the predictive genetic testing of healthy children for such disorders. Of course the family (and perhaps the child if sufficiently mature) may be relieved if a favourable result is given *but* the result will often not be favourable, and even if it is the child may still become a casualty of testing.

PARENTAL RIGHTS

Surely, it may be said, it is a parent's right to insist upon testing their child, both because the parent may simply wish to know the child's genetic status and because the parent may be thought to have a duty to discover any fact relevant to the child's present or future welfare. A child's parents will usually have the best interests of their child at heart, and there is no reason to equate the seeking of genetic testing for a child with child abuse; so why not comply with the wishes of the parent? While the concept of a parental 'right to know' has been argued in North America [18,19], although not by all parties [20], this approach has been discarded in Britain. The Children Act of 1989 considers parents to have *duties* towards their children rather than *rights* over them [21,22]. Decisions about the child's healthcare – including predictive genetic testing – would need to be made on the basis of the long-term best interests of the child. A health professional who considered that genetic testing would run counter to the child's interests, and should be deferred until the child is older, could therefore legitimately refuse to carry out such tests.

While the process of genetic counselling will usually resolve differences of opinion between professionals and parents, this will not always happen. If family and professionals disagree, whose opinion should prevail? In the short-term it may be the professional who 'wins', but this is not likely to be the end of the matter. Except in extreme cases, it will usually be possible for the family to find another professional who is willing to comply with their request. If not, it may be possible for the family to obtain the testing, albeit under false pretences, through commercial channels – by sending a sample for analysis through the post, as if the sample came from an adult. This reinforces the importance of a sensitive and non-confrontational counselling process in advance of any decision about testing, which will usually lead to a consensus view on how to proceed.

WHO IS A 'CHILD'?

The discussion so far has begged the question of who is to count as a 'child'. It is clear that any young child of preschool age will be unable to consider the full range of issues involved in genetic testing. It is equally clear that some 17-year-old individuals are as emotionally and intellectually mature as some 18-year-olds, or 15-year-olds as 16-year-olds. It would not be helpful to adopt a sharp age-related criterion to draw a line between childhood and adulthood. This would inevitably be arbitrary and indefensible.

It will be much more helpful to treat each child and each set of family circumstances as individual. It is recognized in British law that children under 16 years can be competent to make decisions about their health care and reproduction without parental involvement, depending upon the maturity of the 'child'. What will be important in the context of genetic testing is to examine three factors. First, it will be necessary to ensure that a request for testing a legal minor comes from the minor and not from other individuals or agencies. Second, it will be necessary to ensure that the minor is indeed sufficiently mature to appreciate the possible consequences of a favourable or an unfavourable test result, including the consequences for reproduction, career, insurance, self-esteem and relationships within the family. Thirdly, it will be important to find out why the request for testing has come now rather than in a few years time, and why a test should be performed *now*, at this time rather than another. Given a satisfactory discussion and clarification on these points, genetic testing of minors may be appropriate.

An assessment approach has been developed for handling requests from adolescents for predictive testing for HD [23]. Work with children in other health care contexts, especially consent for surgery, has shown that the maturity of a child can be greatly influenced by relevant experience, and that chronological age is not an adequate guide for assessing competence to participate in decision-making [24]. In this chapter, therefore, when caution is expressed in relation to genetic testing in childhood, this does not imply the adherence to an arbitrary age-based definition of childhood; nor does it imply that competence is suddenly acquired or thrust upon a 'child' as they become a 'mature adult'. Children acquire the ability to contribute to the making of important decisions at different rates, and the extent to which a particular child can participate in a specific decision will depend upon many individual and context-specific factors that have to be considered in the counselling before any decision is made about testing in that case. The focus of this chapter, then, is on situations where all would agree that the child is too young to participate seriously in the making of an important decision. This defines an approach which can then be modified appropriately in the various grey areas that arise in practice.

ADJUSTMENT TO GENETIC INFORMATION

It can be argued that a child will be able to adjust more readily to unwelcome genetic information than will an adolescent or young adult. Unfortunately, there is very little evidence on this point. We need more information about the different ways in which families have communicated information about genetic risk and genetic tests, and how the young people concerned have responded to the information. Indeed, it may be true instead that adolescents respond better to genetic information and test results if they are explicitly given control over the process of counselling and testing instead of simply being presented with the facts. This may serve to reinforce their self-esteem and to promote their coping abilities.

CARRIER TESTING

The arguments deployed above could also be applied to the rather different context of the genetic testing of children to identify unaffected carriers of genetic disease, especially autosomal or sex-linked recessive diseases but also carriers of balanced, familial chromosomal rearrangements. In these circumstances, the test result would not be relevant to the health of the child at all, but could be relevant to the child's future reproductive decisions. A test result indicating that the child carried one of these conditions would mean that their future children would be at risk of being affected. In the case of an autosomal recessive disorder, the child's future children would be at risk if their partner was also a carrier of the same condition. If the child was a carrier for a sex-linked disease or for a balanced chromosomal translocation, then any of her future children would be at risk irrespective of the genetic constitution of her partner.

The arguments against carrier testing in childhood are similar to those given above in the context of predictive testing, but they are somewhat weaker. The abrogation of autonomy and confidentiality will occur in this context too, but might not be viewed as so serious as in predictive testing. Testing may result in alterations to the parents' expectations of the child, but their scope will be more restricted – being focused on issues of personal relationships and future reproduction. The other issues of genetic discrimination and the stigmatization of carriers are real issues that must not be forgotten, because examples of such consequences have arisen in a number of different carrier screening programmes, but they are discussed in the chapter on screening for carriers of recessive disorders and they will not be rehearsed here.

In families where a condition is known to be transmitted through healthy female carriers, such as the sex-linked disease Duchenne muscular dystrophy, it is clear that girls are often brought up with powerful expectations about the pattern of their future relationships and reproduction being held by parents (especially by their mothers), and these expectations are frequently adopted by the daughters [25–27]. Such patterns of family expectation arose before genetic testing was able to distinguish between those who did carry the disease in question and those who did not. With the development of accurate methods of carrier testing, families may consider clarification of a child's carrier status while she is still too young to be involved in the discussion. Is it reasonable to comply with such requests, or should the professionals act as gatekeepers to testing and be more cautious, seeking to protect the child's future rights? In autosomal recessive diseases such as cystic fibrosis, it is also known that the healthy sibs of affected children have many needs for information, counselling and support whether or not they turn out to be carriers [28,29]. In some families there are strong family expectations about genetic testing. Once again, should professionals comply with requests to test young children?

As in the context of predictive testing, an open discussion with the parent(s) or others requesting testing – reassuring them that testing will be available when the request for it comes from the child and at a time when it is relevant to the child – will often resolve the issue. The parents may have thought of testing the child as a way of discharging the duty of responsible parents to ensure that the child is tested. In fact, of course, testing a young child does not discharge the duties of either the parents or the professionals, it merely alters them. Whatever the test result – whether the child is a carrier or not – it will still be necessary for the parents to raise with the child the possibility of their being a carrier at an age when they can understand the significance of the question. Unless the child is forever kept unaware of the genetic condition that has occurred in the family and so is unaware of their potential carrier status, even a negative test result (not a carrier) will have to be explained to the child when older. The professionals involved with testing a young child for carrier status will also not thereby have discharged their duties simply by giving the result to the family; there remains an obligation to ensure that the child is given the opportunity to discuss their result when older, thereby ensuring that they actually are given the result and that they understand its significance. It is all too easy for families to forget about such tests (especially if a crucial family member dies or becomes ill), or to misunderstand the significance of a test result.

Some families will express a very strong desire to test a child for carrier status despite a very full discussion. There may be a hidden agenda in some such cases, such as a desire to test paternity, but it may simply be a strong 'desire to know'. The family context will often be of parents who have lost a brother, sister, son or daughter with a serious and most distressing disease; unresolved grief in the parents may, very understandably, intrude into the decisions being made about the child. In my experience, it is not helpful for professionals to refuse to perform carrier tests under these circumstances. It is very reasonable to explore the motivation and reasoning underlying the parents' request, to indicate that testing will not resolve all the issues even if the child is shown not to be a carrier and to explain the possible difficulties that may arise from testing in childhood; but if the parents continue to insist on the child being tested then it may be appropriate to agree to this, because a persistent refusal generates antagonisms that may block the resolution of whatever factors are underlying the request. One approach to handling this would be to agree to test the child in principle but to insist upon a delay of at least a few weeks, after which the family has to re-initiate contact with the clinic; the decision is put entirely in their hands. This may give the family a chance to reassess their decision without the fear that the test will be denied them, so that for the first time they become free to examine all the factors for and against testing and then to make up their own minds.

SUSCEPTIBILITY SCREENING FOR COMMON DISEASES

Experience with screening children for genetic susceptibility to common diseases is limited, but the experience that has been accumulated amounts to a cautionary tale. A report from Bergman and Stamm some 30 years ago, introduced the term 'cardiac non-disease' to describe the emotional and physical effects of inappropriately labelling healthy children who had benign cardiac murmurs as being sick [30]. In the newborn period, a Swedish programme of screening to identify infants affected by α_1-antitrypsin deficiency was stopped because of damaging social consequences. Affected children are more susceptible to lung disease in adult life and should be protected from exposure to cigarette smoke, but the fathers of the identified infants actually smoked more heavily than the fathers of control infants [31]. In USA, children found to have above-average serum levels of cholesterol have sometimes been subjected to excessively restrictive diets, even leading to growth failure from malnutrition [32,33], and sometimes their possible special health needs have been ignored [34].

The lessons I draw from these reports are: (i) that programmes of screening for genetic disease in childhood should be piloted and considered very carefully before being adopted on a wide scale, and (ii) that labelling healthy children as having a faulty gene could well lead to similar problems to those reported by Bergman and Stamm, and should not ever be undertaken lightly.

AN 'IDEAL'(?) APPROACH TO GENETIC INFORMATION IN CHILDHOOD: THE RELEVANCE OF ADOPTION

While requests to carry out predictive genetic tests for late-onset diseases on healthy children rouse fears in many professionals that such testing will remove the child's future rights to autonomy and confidentiality and may be damaging to the child's welfare, these fears are not so marked in relation to carrier testing of purely reproductive significance to the future adult. As discussed above, therefore, it may be reasonable to accede to a family's request for carrier testing after a full exploration of the reasons underlying the request and after counselling about the possible implications of the testing for the child; the consequences for the emotional welfare of the family of a persistent refusal to test may outweigh the reasonable concerns about testing. For predictive testing, in contrast, my judgement is that the protection of the child's future rights and interests will permit their testing only in the most exceptional circumstances. The preferred course of action in both circumstances, relating to predictive and carrier testing, would be to offer full counselling and to defer the testing until the child could participate and make up his or her own mind when older. So when a family requests genetic testing of their child, what would be the optimal set of arrangements to make with them in relation to genetic testing for the child in the future?

A guide to the best course of action in these circumstances may come from the general advice offered in relation to telling adopted children about their biological parents. It would be a mistake ever to lie to or to mislead children, and the best policy is to give information to children at the rate at which they seek it. In this sense, it has much in common with 'breaking bad news' in the context of cancer diagnoses [35]. Children can be introduced to the idea that they have been adopted – indeed, they have been 'specially chosen' – from a very early age. But they are not usually given much information about their biological parents until they are older. Adopted children are not permitted to seek out their biological parents until they are adult. A similar approach can be used with genetic information; children can be brought up in the knowledge that they may share certain genes with other family members, and that they will be able to find out more about this once they are older. The parallel is very close, and of course genes are involved in both situations.

If both the parents and the professionals are to be reassured that they are discharging their responsibilities properly, then there is clearly a need for someone to ensure that the child is offered genetic counselling at an appropriate time in the future. The family may 'forget' about the genetic disease, especially if the affected individual in the family died before the 'child' was born. This is particularly likely to occur if a parent or another crucial family co-ordinator dies, or if the parents separate or divorce. But there is often no clear mechanism for the health professionals to 'remember' to contact the child when older, and in any case the family may well have moved house or country in the meantime.

This is not an argument for testing children in childhood, because the need for genetic counselling remains even once the test has been carried out – it could even be seen as more important because the family may have misunderstandings and may not appreciate that further discussion is required if they think that the testing process has been completed. Even if the family was perfectly informed 10 years previously, testing and the interpretation of test results may have developed in the meantime. Rather, this is an argument for the continued existence of regional genetic counselling services. These services should be adequately supported so that they can maintain infrequent but regular contact with families in which individuals may wish in the future to be offered genetic counselling and perhaps testing. This amounts to an argument for the funding of genetic services to enable them to maintain genetic registers – lists of patients who can be offered contact in the future for genetic counselling or, in a wider context, for surveillance for the possible complications of genetic disorders.

At the present, the transmission of genetic information (knowledge about a disease that has cropped up in the family, as well as of test results and the appropriateness of genetic counselling for other family members in the future) relies upon three systems: family knowledge and willingness to transmit relevant information, the general medical practitioner or 'family doctor', and genetic counselling services. With the increasing complexity of genetic information, the increase in family instability and the more frequent changes in 'family doctor' that accompany social mobility, there is a good argument for strengthening the role of genetic services to compensate for changes in the other two systems. This is not a call for a new institutional framework, it is a plea for the more adequate support of genetics services within the health service to permit them to carry out more systematically, a role that has often been thrust upon them and for which they have never been adequately resourced. This type of genetic register would be completely voluntary and would require the active co-operation of the families for it to function; it is not an attempt to impose a new legal burden on genetic services never to lose contact with their clients [36].

EVIDENCE AND THE FUTURE

The Clinical Genetics Society in Britain established a working party to examine the range of attitudes and practices in relation to the genetic testing of children, and it examined in particular the problematic areas of predictive and carrier testing in childhood. The Working Party reported in 1994 [37], and emphasized the great diversity of attitudes and practices both within and between various professional groups. In general terms, it was apparent that genetic nurses and co-workers were more 'protective' of children than were the clinical geneticists, who were themselves more 'protective' than paediatricians or other medical specialists. The geneticists were more selective in the tests that they did or would perform on children, whereas the paediatricians expressed their willingness to comply with parental requests in most settings. The interpretation of these findings by the Working Party was that most paediatricians had not been faced by these issues very much in practice, while the geneticists had had more opportunity to reflect upon difficult cases and adverse outcomes of a range of counselling scenarios.

Although many professionals and other groups recommend caution in relation to genetic testing in childhood, there is little systematic evidence of it causing any harm. This is scarcely surprising because such testing has only recently become possible for most conditions, and there has been an understandable reluctance to offer predictive tests to children unless there is some medical benefit to be gained – this reluctance perhaps being one manifestation of 'appropriate paternalism'. The absence of evidence that such testing causes harm is not a reason for going ahead with such testing, particularly when there are good grounds for caution. Not only is there the abrogation of the child's rights to autonomy and confidentiality to consider, and the disadvantages of discrimination and stigmatization that face adults with unfavourable genetic test results [37] (in relation to both predictive tests and tests of carrier status) but there is also the question of the impact of test results on the emotional and social development of the child. Given this lack of evidence, there has so far been a considerable degree of consensus among published reports and guidelines in relation to the predictive testing of children for late-onset disorders, recommending great caution [38–41]. The Genetics Interest Group, an umbrella organization of lay support groups in Britain, responded critically to the Clinical Genetics Society report, but they opposed such testing even more firmly [17], although adopting a more liberal approach to carrier testing.

The question of evidence, or the lack of it, is primarily relevant in relation to the social and emotional consequences of testing; it has much less to say about the abrogation of rights, and to what extent that is a matter for concern. What is required now is the careful study of family and professional strategies for transmitting genetic information, including assessments of the success of various strategies in different family circumstances. It would be very unhelpful at this stage of our understanding simply to carry out widespread genetic testing of children and to wait the many years necessary for evidence of harm to accumulate. It would not be feasible to mount a randomized, controlled trial of two different approaches because there are so many dimensions to this issue that it could not meaningfully be reduced to a simple dichotomy. It will be most helpful now to record current practice systematically, and to base advice to families in a broad range of different contexts on the analysis of this experience.

While 'evidence' is important and helpful both in arriving at general policy and in making decisions about individual cases, it would be unwise to ignore the equally important roles of ethical reflection and conceptual clarification. Calls for more evidence and for further empirical studies do chime with the times in this era of evidence-based medicine, but it is all too frequently forgotten that 'facts' do not exist independently of the observer but are constructed socially by individuals and groups who have explicit opinions, implicit attitudes and vested interests which will shape these 'facts'. The pose of being a neutral, objective observer of one's fellow humans, able to discover 'the truth' about some aspect of mankind, has a long history but has been intellectually untenable for the best part of a century.

A second area in which research will be helpful, as pointed out by Michie and Marteau [42] as well as Alderson [24], is the maturation of the child's competence to participate in decision-making. This may help us to make more appropriate judgements in that grey area where the 'child' is able to make some contribution to important decisions about genetic testing.

Finally, I will add that there has been considerable debate among professionals about childhood genetic testing over the past few years [43–49]. That the issue has aroused such interest is a very positive development, because it used to be neglected; it was an invisible problem. On the other hand, it is disappointing that some parties have tended to polarize the debate in an oversimplistic fashion, portraying one camp as opposed to testing and another as for it. Let us hope that the next few years sees the debate widen to include other standpoints, and that professionals and families can move forward together, towards a common approach to the management of these difficult issues.

REFERENCES

1. Craufurd D, Harris R (1986) Ethics of predictive testing for Huntington's chorea: the need for more information. *Br. Med. J.* **293**: 249–251.
2. World Federation of Neurology: Research Committee Research Group Ethical issues policy statement on Huntington's disease molecular genetics predictive test. *J. Neurol. Sci.* (1989) **94**: 327–332 and *J. Med. Genet.* (1990) **27**: 34–38.
3. Bloch M, Hayden MR (1990) Opinion: predictive testing for Huntington disease in childhood: challenges and implications. *Am. J. Hum. Genet.* **46**: 1–4.
4. Kessler S (1988) Invited essay on the psychological aspects of genetic counselling. V. Preselection: a family coping strategy in Huntington Disease. *Am. J. Med. Genet.* **31**: 617–621.
5. European Community Huntington's Disease Collaborative Study Group (1993) Ethical and social issues in presymptomatic testing for Huntington's disease: a European Community collaborative study. *J. Med. Genet.* **30**: 1028–1035.
6. Codori A-M, Brandt J (1994) Psychological costs and benefits of predictive testing for Huntington's disease. *Am. J. Med. Genet.* (*Neuropsych. Genet.*) **54**: 174–184.
7. Huggins M, Bloch M, Wiggins S *et al.* (1992) Predictive testing for Huntington Disease in Canada: adverse effects and unexpected results in those receiving a decreased risk. *Am. J. Med. Genet.* **42**: 508–515.
8. Lawson K, Wiggins S, Green T *et al.* (1996) Adverse psychological events occurring in the first year after predictive testing for Huntington's disease. *J. Med. Genet.* **33**: 856–862.
9. Tibben A, Vegter-van der Vlis M, Skraastad MI *et al.* (1992) DNA-testing for Huntington's disease in The Netherlands: a retrospective study on psychosocial effects. *Am. J. Med. Genet.* **44**: 94–99.
10. Ball D, Tyler A, Harper PS (1994) Predictive testing in adults and children. In: *Genetic Counselling* (ed. A. Clarke) pp. 63–94. Routledge, London.
11. Craufurd D, Dodge A, Kerzin-Storrar L, Harris R (1989) Uptake of presymptomatic testing for Huntington's disease. *Lancet*, **ii**: 603–605.
12. Bloch M, Adam S, Wiggins S, Huggins M, Hayden MR (1992) Predictive testing for Huntington disease in Canada: the experience of those receiving an increased risk. *Am. J. Med. Genet.* **42**: 499–507.
13. Tyler A, Morris M, Lazarou L, Meredith L, Myring J, Harper PS (1992) Presymptomatic testing for Huntington's disease in Wales 1987–1990. *Br. J. Psychiat.* **161**: 481–489.
14. Morris M, Tyler A, Harper PS (1988) Adoption and genetic prediction for Huntington's disease. *Lancet*, **ii**: 1069–1070.

15. Morris M, Tyler A, Lazarou L, Meredith L, Harper PS (1989) Problems in genetic prediction for Huntington's disease. *Lancet*, **ii**: 601–603.
16. Harper PS, Clarke A (1990) Should we test children for 'adult' genetic diseases? *Lancet*, **335**: 1205–1206.
17. Dalby S (1995) Genetics Interest Group response to the UK Clinical Genetics Society report 'The genetic testing of children'. *J. Med. Genet.* **32**: 490–491.
18. Pelias MZ (1991) Duty to disclose in medical genetics: a legal perspective. *Am. J. Med. Genet.* **39**: 347–354.
19. Sharpe NF (1993) Presymptomatic testing for Huntington Disease: is there a duty to test those under the age of eighteen years? *Am. J. Med. Genet.* **46**: 250–253.
20. Clayton EW (1995) Removing the shadow of the law from the debate about genetic testing of children. *Am. J. Med. Genet.* **57**: 630–634.
21. Montgomery J (1993) Consent to health care for children. *J. Child Law*, **5**: 117–124.
22. Montgomery J (1994) Rights and interests of children and those with mental handicap. In: *Genetic Counselling* (ed. A. Clarke) pp. 208–222. Routledge, London.
23. Binedell J, Soldan JR, Scourfield J, Harper PS (1996) Huntington's disease predictive testing: the case for an assessment approach to requests from adolescents. *J. Med. Genet.* **33**: 912–918.
24. Alderson P (1993) *Children's Consent to Surgery*. Open University Press, Buckingham.
25. Parsons EP, Atkinson P (1992) Lay constructions of genetic risk. *Sociol. Hlth Illness*, **14**: 437–455.
26. Parsons EP, Atkinson P (1993) Genetic risk and reproduction. *Sociol. Rev.* **41**: 679–706.
27. Parsons EP, Clarke A (1993) Genetic risk: women's understanding of carrier risk in Duchenne muscular dystrophy. *J. Med. Genet.* **30**: 562–566.
28. Fanos JH, Johnson JP (1995) Perception of carrier status by cystic fibrosis siblings. *Am. J. Hum. Genet.* **57**: 431–438.
29. Fanos JH, Johnson JP (1995) Barriers to carrier testing for adult cystic fibrosis sibs: the importance of not knowing. *Am. J. Med. Genet.* **59**: 85–91.
30. Bergman AB, Stamm SJ (1967) The morbidity of cardiac nondisease in schoolchildren. *New Engl. J. Med.* **276**: 1008–1013.
31. Thelin T, McNeil TF, Aspegren-Jansson E, Sveger T (1985) Psychological consequences of neonatal screening for alpha-1-antitrypsin deficiency. *Acta Paediatr. Scand.* **74**: 787–793.
32. Lifshitz F, Moses N (1989) Growth failure. A complication of dietary treatment of hypercholesterolemia. *Am. J. Dis. Child.* **143**: 537–542.
33. Newman RT, Browner WD, Hulley SB (1990) The case against childhood cholesterol screening. *J. Am. Med. Assoc.* **264**: 3003–3005.
34. Bachman RP, Schoen EJ, Stembridge A, Jurecki ER, Imagire RS (1993) Compliance with childhood cholesterol screening among members of a prepaid health plan. *Am. J. Dis. Child.* **147**: 382–385.
35. Buckman R (1992) *How To Break Bad News: a Guide for Health-care Professionals*. Papermac, London.
36. Hoffmann DE, Wulfsberg EA (1995) Testing children for genetic predispositions: is it in their interest? *J. Law Med. Ethics*, **23**: 331–344.
37. Billings PR, Kohn MA, de Cuevas M, Beckwith J, Alper JS, Natowicz MR (1992) Discrimination as a consequence of genetic testing. *Am. J. Hum. Genet.* **30**: 476–482.

38. Working Party of the Clinical Genetics Society: Report on the Genetic Testing of Children. *Clin. Genet. Soc.* (1994) and *J. Med. Genet.* (1994) **31**: 785–797.
39. Wertz DC, Fanos JH, Reilly PR (1994) Genetic testing for children and adolescents: who decides? *J. Am. Med. Assoc.* **272**: 875–881.
40. American Society of Human Genetics/American College of Medical Genetics Report (1995) Points to consider: ethical, legal and psychological implications of genetic testing in children and adolescents. *Am. J. Hum. Genet.* **57**: 1233–1241.
41. British Paediatric Association Ethics Advisory Committee (1996) *Testing Children for Late Onset of Genetic Disorders*. British Paediatric Association, London.
42. Michie S, Marteau TM (1996) Predictive genetic testing in children: the need for psychological research. *Br. J. Hlth Psychol.* **1**: 3–16.
43. Clarke A, Flinter F (1996) The genetic testing of children: a clinical perspective. In: *The Troubled Helix: Social and Psychological Implications of the New Human Genetics* (eds T.M. Marteau and M.P. Richards) pp. 164–167. Cambridge University Press, Cambridge.
44. Michie S (1996) Predictive genetic testing in children: paternalism or empiricism? In: *The Troubled Helix: Social and Psychological Implications of the New Human Genetics* (eds T.M. Marteau and M.P. Richards) pp. 177–183. Cambridge University Press, Cambridge.
45. Michie S, McDonald V, Bobrow M, McKeown C, Marteau T (1996) Parents' responses to predictive genetic testing in their children: report of a single case study. *J. Med. Genet.* **33**: 313–318.
46. Clarke A (1997) Parents' responses to predictive genetic testing in their children. *J. Med. Genet.* **34**: 174.
47. Clarke A, Harper PS (1992) Genetic testing for hypertrophic cardiomyopathy. *New. Engl. J. Med.* **327**: 1175.
48. Harper PS, Clarke A (1993) Screening for hypertrophic cardiomyopathy. *Br. Med. J.* **306**: 859–860.
49. Chapple A, May C, Campion P (1996) Predictive and carrier testing of children: professional dilemmas for clinical geneticists. *Eur. J. Genet. Society* **2**: 28–38.

Chapter 3

PRESYMPTOMATIC TESTING FOR LATE-ONSET GENETIC DISORDERS

Lessons from Huntington's disease

P.S. Harper

INTRODUCTION

The feature that particularly distinguishes DNA-based testing for inherited disorders from all other kinds of test is its constancy and its separation from the actual occurrence of disease. Any change detected in DNA due to a germline mutational change is likely to be present essentially unchanged from conception to death, unaltered by whether the disease state is present or not. The extent or nature of the genetic change may show correlation with the likely severity or age at onset of disease, but in most cases, there is no accurate guide to prognosis in an individual. Thus, for late-onset genetic disorders, DNA based genetic testing in a healthy person at risk will create knowledge of likely future disease after an interval possibly of many years, during which time the person will remain entirely healthy – true presymptomatic detection.

The differences between these genetic tests and other approaches to early diagnosis are so great that most people, including those undertaking such tests and those being tested, have not yet adjusted to them. Imaging techniques can often show minor structural changes presymptomatically in such disorders as polycystic kidney disease, but these are age related and progressive. Enzyme measurements, which are relatively constant, have been used principally in childhood metabolic disease, abnormality implying disease onset in the near future. The result is that those individuals receiving an abnormal result are considered by themselves and others to 'have the disease', even though still presymptomatic.

The advent of DNA based testing across a wide and increasing range of late-onset genetic disorders, where presymptomatic detection can imply an interval of 30 years or more before onset, is a challenge to conventional thinking about medical tests. Indeed, as mentioned in Chapter 1, it can be questioned whether such testing is medical at all, an issue which is of particular relevance to the field of insurance (see Chapter 4), where companies currently expect 'medical information' to be declared. Regardless of this, the delivery of a service of presymptomatic genetic testing for late-onset diseases creates a series of challenges if we are to avoid inappropriate use and serious harm to those involved. Some of the issues involved are discussed here in the context of Huntington's disease (HD), using my own experience of this disorder and that of others internationally.

WHY USE HUNTINGTON'S DISEASE AS A MODEL?

When one has worked on a particular disorder for 25 years, as I have in the case of HD, it is natural to be convinced that the condition must necessarily be important and suitable for drawing wider conclusions. The choice of HD as a model thus needs to be justified objectively, and *Table 3.1* lists some of the main criteria that make it particularly suitable. As a progressive brain disease, giving a combination of mental deterioration with serious physical disability, notably involuntary movements, it is a condition that creates a severe burden for affected families [1]; while relatively rare (around 1 in 10000 prevalence) as a disease, the frequency of those at high risk as first-degree relatives is much greater (around 1 in 2000) not including spouses and more distant family members.

Although it was only in 1993 that a specific genetic mutation for HD was identified [2], the gene locus was one of the earliest to be mapped [3], allowing genetic testing by linked markers to have been offered for the past 10 years. This early start, coupled with the effort of all workers involved to co-ordinate and document their approaches [4,5], has meant that it has been possible to share and compare experience on a world-wide basis [6] in a way that has been the case for few other diseases. I am tempted to add that another advantage is that presymptomatic testing has mainly involved genetics centres who have

Table 3.1. Huntington's disease as a model for presymptomatic testing for late-onset genetic disorders

Serious and ultimately fatal
Currently not treatable
Onset most often in middle life
Affects successive generations (autosomal dominant)
Relatively frequent
Specific genetic testing feasible
Testing introduced with careful preparation
Accurate documentation of testing experience
Close co-operation and co-ordination of protocols world-wide

viewed the activity as prediction, rather than as a form of early diagnosis, and have been accustomed to dealing with the family issues involved.

Molecular genetic testing for HD has a particular added advantage in comparison with many other genetic disorders in that the mutational basis is a single, though variable, genetic change in virtually all patients world-wide [7]. Thus, one is not confused by heterogeneity or geographical variation as is so often the case for other inherited diseases.

DEFINING THE ISSUES AND PROBLEMS

Long before any form of genetic testing became possible, HD was well recognized as producing a series of particularly difficult issues in genetic counselling. Indeed, I strongly suspect that this was, and continues to be one of the main reasons why neurologists and other clinicians have been only too happy to leave this field to geneticists! The great variability in age at onset, the problems caused by mental symptoms, the associated social difficulties and the need for seeing large and often scattered extended families, has made the condition particularly challenging, especially if, as in our own centre, an effort has been made to provide a service on a broad population basis [8].

The potential pitfalls and dangers of presymptomatic testing for HD were widely recognized before anyone started to offer it as a service, and were the main reasons why the initial programmes built-in careful preparation, support, and evaluation. Our own centre concentrated from the outset in identifying and analysing the problems encountered [9]. As I rationalized the situation at the time for my somewhat sceptical colleagues, since we were bound to meet the problems anyway, we might as well make them the principal focus of our work, and help others to learn from them. During the subsequent decade, we have certainly had no shortage of problems, and it has become increasingly clear that most of them are thoroughly relevant to other forms of late-onset genetic disease for which genetic prediction is becoming possible.

Table 3.2 summarizes the main categories of the issues and problems; they can broadly be divided into four main categories, and it is worth considering all of them in turn. In most cases, it will be self-evident that they apply to other disorders too, but the general aspects will be returned to, later in the chapter. Some of the examples given have formed part of a four-centre European study, which found the issues to be remarkably similar in the different countries [10].

PROBLEMS OF CONSENT

The principle of informed consent is a cornerstone of medical practice, but one that has been widely ignored in the context of medical investigations, particularly if these are not in themselves dangerous, even though their consequences may be profound. Genetic testing falls into this category; while its use in a

Table 3.2. Issues in presymptomatic testing for Huntington's disease

Problems of consent
Referral for testing without consent
Testing of children
Referral under third party pressure
Problems involving relatives
Privacy of information
Refusal to supply information or sample
Unsolicited risk alteration; testing at 25% risk
Laboratory issues
Duplicate samples
Use of research samples
Pseudonyms and anonymous requests
Issues of confidentiality
Refusal of disclosure to family doctor
Refusal of disclosure to spouse or relative
Third party requests for information
Disclosure to insurance companies or employers

research context will be regulated by ethical review committee approval, and a consent form generally required, no such control currently exists for service use, where genetic testing has often been considered as another category of medical investigation, without regard to its wider consequences. This is reinforced by the fact that in many disorders, including HD, the same test may be used both diagnostically, in a person showing clinical symptoms, and as a prediction in a totally healthy individual. The ease and danger of confusing diagnostic and predictive testing has already been discussed in Chapter 1.

From the outset, all research protocols for HD presymptomatic testing insisted on written consent, and this has continued in service practice. No problems have been reported with this, and given the consequences of the test result, which are at least as significant as any surgical operation or procedure, the practice seems eminently sensible. In the course of our own experience of HD predictive testing, involving over 300 applicants, we have encountered a number of instances where consent was an issue, which are worth discussing in some detail. Some of these were reported in the European study [10].

The first situation is where testing has been requested without the individual's knowledge or permission, as in the following examples. Of seven such instances encountered in our own series, five were from psychiatrists. The particular issues involving predictive testing in psychiatric practice have been discussed by Scourfield *et al.* [11].

Example 1

A psychiatrist referred a 50-year-old man for testing. He had been under psychiatric care for many years, suffering from severe depression but had never showed any neurological signs. It was suspected that the depression was related to the family history of HD. The psychiatrist asked for the patient to be tested without his knowledge and offered to send a blood sample. He also asked for the genetic centre to send the result to him (the psychiatrist) so that he could decide whether or not to inform this patient. The genetic centre explained why this was not possible and obtained permission to see the patient, who was quite disinterested in testing; therefore, it was not carried out.

Example 2

A lawyer asked for a 24-year-old man at risk for HD to be tested, in connection with a charge of sexual assault, without his permission or knowledge. The man was suffering from acute mental illness at that time but, apparently, not displaying any neurological or dementing signs. The genetic centre declined the lawyer's request but offered to see the patient, when recovered, if he wished it. There has been no subsequent contact.

This second example illustrates another issue involving consent: where the individual to be tested is mentally incapable; because of acute illness in this particular case, but mental handicap could be another reason. Because HD may occasionally present with an acute psychosis, testing may be requested in such circumstances by psychiatrists, something we have always refused. The potentially serious consequences are obvious; following recovery from the episode the individual might be faced with the knowledge and long-term consequences of having the HD mutation, having never consented and possibly never wished for this.

An important group of individuals unable to give full informed consent are children, especially young children. It was the experience of ourselves and others in receiving requests for predictive testing on such young children, usually from a parent, but also from social workers and adoption agencies [12], that first raised the general issue of testing children for 'adult' genetic disorders, a topic more fully explored in Chapter 2. This has provided a striking example of the general applications of our HD experience, with the issues clearly defined and identified early for HD, but now recognized as relevant to a wide range of other genetic diseases.

Although these issues of consent are now widely appreciated, at least by geneticists, it is likely that they will become an increasing problem in all late-onset disorders as genetic testing becomes more widely used by clinicians who may not give thought to the need for consent. In particular the simplification of technology and development of batteries of tests for multiple disorders, could mean that numerous people are tested without consent for disorders of which they are not even aware. This is an issue to which government bodies and others involved in the regulation of genetic testing need to give urgent consideration.

Allied to consent is the issue of third party pressure, which can be very difficult to assess, even with full discussion. Often the spouse or partner of the person at risk is clearly the one who has initiated a request for testing, and the individual to be tested may be reticent or uncommunicative. Where the situation is more overt, as in the example below, it is usually possible to resolve it reasonably.

Example

Three siblings in their early twenties were referred by their mother who was already looking after their father with advanced HD. She wanted them to be tested so that she could make plans for the future, knowing how many would need her care. It was explained to her that her children must make up their own minds about testing. Two joint family interviews were held, after which separate interviews were arranged for the children. Two out of the three eventually, of their own free will, decided to be tested.

Ultimately, the definition of whether a person is acting of their own free will is perhaps a philosophical one, and in the practical context of presymptomatic testing, one can only do one's best to ensure that the person tested does have the full information needed for a decision to be reached.

ISSUES INVOLVING RELATIVES

One of the defining characteristics of genetic testing is that it has implications for family members; indeed this is frequently the main reason why it is requested in the first place. If this feature is not recognized, however, serious problems can arise; our HD experience has illustrated several of these, not all of them expected.

Until the HD mutation was identified, predictive testing was based on linked gene markers and a prediction could only be made if samples were available from a number of family members, to allow the pattern of transmission in the family to be deduced. This meant that many people wishing for testing could not be offered it, because key members were dead or unavailable. It also created difficulties on account of the need for family members to be approached, as seen in the examples given below. Although such problems are now less common in HD prediction, they are highly relevant to testing for other genetic disorders where specific mutations have yet to be identified.

Example 1

A 35-year-old lady was distressed to receive what she described as a 'peremptory' request from her sister, living in another part of the UK, for a blood sample. The woman had applied for testing and had, apparently, been asked by her local genetic centre to obtain this sample. Her sister was worried and very depressed about marital problems and vague non-specific symptoms which, she felt, could herald the onset of HD. She had firmly decided not to be tested as she felt that an adverse result could cause her to become suicidal. She was reassured that she did not need to provide a blood sample and was offered support, if needed, in any further communications with her sister.

Example 2

An affected family member, living in another part of the country, was approached by his nephew for a blood sample. He was willing to provide one, but his family doctor was unwilling to take it because he had acquired immune deficiency syndrome (AIDS). The genetic centre did not disclose to the applicant the reason why the blood sample was unobtainable.

Example 3

A 22-year-old, divorced, unemployed man who was living at home, applied for a test. He came from a very small family but testing by genetic linkage was possible if the diagnosis could be confirmed in an affected sibling (known to the genetic centre but not to the applicant, his family or to the sibling himself). It was considered unethical for the genetic centre to approach the sibling and, therefore, testing was postponed until the diagnosis became known in the family.

These examples show how unpredictable these problems are, since neither the person requesting testing nor the testing centre may be aware of the background complexities, while the person contacted may know little or nothing about predictive testing or even about the family disorder. The last example shows that it is not just the obtaining of a sample that can create problems, but privacy of other information. A genetic centre will frequently know that a family member is affected with HD at a stage when relatives are not aware of this, or when even the affected individual may not have accepted their own diagnosis.

If a relative refuses to provide a blood sample, then the situation is normally straightforward, even if unsatisfactory. Occasionally, however, colleagues trying to be helpful can compound the situation, as in the following example.

Example 4

A man in the middle stages of HD refused to give blood on the grounds that there was no such disease in his family. He was estranged from his children, who were the applicants for testing, after he had married, very recently, for the second time. The applicants felt angry that their father had 'let them down again' and successfully prevailed on the patient's general practitioner to obtain a blood sample without telling him of the use to which it was going to be put. The genetic centre then had to decide whether it was ethical to use the sample. Fortunately, in the meantime another affected relative had been identified, so that the ethically dubious sample did not have to be used!

A second important issue involving relatives is the possibility that the risk status of others may be altered by the test that is being performed on the applicant. In the past, this was a danger with family-based testing using linked markers, where the typing of a relative to give a result for an applicant might inadvertently alter the risk also for the relative; with specific mutation testing having largely replaced linked markers, this is no longer a problem for HD, but it remains so for other genetic disorders where specific defects have yet to be found.

For HD, the main problem of this type now arises from requests for testing by those at 25% prior risk (i.e. those whose parent remains healthy but still at risk) and where a grandparent has been affected. Clearly the identification of the HD mutation in such a situation will mean that a predictive test will have also been done on the parent, who may not have wished for this; in addition the parent will be likely to be much closer to onset of the disease.

This problematic situation was widely discussed before predictive testing became feasible, and has continued to generate controversy since. There is an ethical conflict between one person's right to know and another's right to privacy. Even in situations where the applicant states that they would not inform their parent, there can be no guarantee that this might not happen. It is possible that this may now be prompting some older individuals to request testing who would not otherwise have done so, on the basis that if they do not, they may find out via their children that they carry the mutation. This provides an interesting counterpart to the more frequent situation of children being dependent on their parents, and illustrates the effects that genetic disorders and genetic information may have on relationships between the generations.

There is no obvious answer to this difficult situation, but most centres involved in testing have found that it occurs rather rarely, largely because it is recognized as an issue, and often resolves after appropriate discussion. This impression has now been confirmed by studies from The Netherlands [12] and UK (UK Huntington's Disease Prediction Consortium, unpublished data), which showed that where testing was requested by those at 25% risk it made up under 10% of all predictions, that the intervening parent was often deceased;

that often either a living parent had agreed to be tested or the person at 25% risk had withdrawn from testing. In the two series, the number of instances where an abnormal result was obtained and where the intervening parent either disagreed with or was unaware of testing of their offspring was minimal: one in around 500 total predictions in each study.

In my view this is a striking example of how serious problems can usually be avoided by careful discussion and provision of information. I strongly suspect that if genetic testing had been implemented for HD as a purely medical or laboratory activity, without the opportunity for full genetic counselling, many more problems of this type would have arisen; it will be extremely important to monitor the situation for other diseases to see if this indeed proves to be the case.

LABORATORY ISSUES

Although the main burden of information-giving and support during presymptomatic testing for HD falls on the clinical and counselling team, it would be wrong to think that the laboratory staff are detached from the whole procedure. Clearly the whole foundations of testing rest on their accurate procedures and interpretation, but there are broader issues of laboratory–clinical links, affecting whether samples are being referred appropriately from outside sources, the form in which reports are drafted, and what should be done when a borderline or otherwise puzzling result is encountered.

Some laboratory-related issues have already been mentioned in the context of samples from relatives – whether refused or obtained inappropriately. An unexpected problem in our own centre arose from an applicant insisting on using a pseudonym, although some samples needed from relatives were sent using the true name, causing considerable confusion. Fears about confidentiality in relation to insurance or employment could favour the use of anonymous requests or pseudonyms, and could increase pressure for 'over the counter' testing (see Chapter 5). I should like to highlight two other laboratory issues where we have found our own practice to vary from that of many other centres and have been criticized for being too strict.

The first concerns the use of research samples, especially those taken from those at risk. We have made it a policy never to type such samples from healthy individuals for specific mutation testing, even though they had previously consented to the sample being given on a research basis. The reasons for our policy are that there is a substantial chance in such a late-onset disorder as HD of generating an abnormal result; also, that one often has no way of knowing whether the person would or would not have received proper counselling and support; and that it would be improper to generate the information and withhold it unless the sample had previously been made anonymous.

This policy has not been followed by all research workers in HD and may well have created considerable problems for those involved, though I have never seen this documented in publications. It is clearly not unique to HD, but will be met in any late-onset disorder where specific mutation testing is possible. I first encountered the issue in relation to prion dementias, and subsequently wrote a short account of the potential dangers more generally to alert colleagues [14]. Despite this, many workers seem still to be remarkably unaware of the issue and regard a consent obtained for a different purpose several years previously as a licence to do anything else on the sample that they wish.

The other HD laboratory issue on which our own centre differs from most others is insisting on using two separate samples taken on separate occasions for all individuals requesting HD presymptomatic testing. The main reason why we implemented this was because, as one of the first centres to undertake testing with linked markers, we were receiving numerous samples on family members by mail, some poorly documented. Also, working in Wales, one frequently has many people sharing the same surname who are being tested and who are totally unrelated. When specific mutation testing began, we reassessed the situation, but concluded that though the likelihood of sample mix-up was now less, it would be much more difficult to detect on a single sample, while the consequence of reversing a result was immense. Since then we have found that the continued use of two separate samples has given great reassurance to both laboratory and clinical staff, and also to those tested, particularly when we have encountered an unexpected result, for example an abnormal result in a person at low risk because of advanced age.

Our policy does of course raise the question: why do we do this for HD but not for all diseases? This is not easily answered, and is relevant when considering the overall implications of using HD as a model for other late-onset genetic disease testing.

ISSUES OF CONFIDENTIALITY

Once a result has been generated, the question arises as to what to do with it. Obviously, the first person to be informed should be the one who has been tested, but we have been asked by clinicians on more than one occasion to give the result only to them since they did not consider that the person tested should be told!

The process of giving the result of a predictive test in HD, especially if abnormal, is always an emotional and often traumatic one for those involved; we always give the result face to face, though some people prefer to have it written down and to be alone, or with a supporter initially. I have described our own experiences elsewhere [15], and others, including those receiving results, have also documented their procedures and reactions [16–20].

The points of confidentiality needing special consideration here relate to wider disclosure of information and are of considerable general importance in terms of testing for genetic disorders overall.

The first type of disclosure is to the family doctor, especially relevant in the UK where everyone is registered with one. Many referrals will have come from or through the family doctor, who also forms the immediate focal point for professional support, or for medical action if symptoms of HD were to arise in the future. On the other hand, people may feel uncertain as to the degree of confidentiality of a family doctor's record. Increasingly, information is shared with other members of the 'primary care team', such as nurses and social workers, while receptionists and clerks may also have access. In small communities this can be of particular concern, and it is not infrequent for applicants to request specifically that their family doctor is not informed. There is also the vexed and unresolved issue, discussed in more detail in Chapter 4, of whether predictive genetic test information on healthy individuals should be released to insurance companies as 'medical information', especially if it has been included in the individual's medical record.

Disclosure to spouses or relatives again raises difficult ethical issues. At a practical level, it can be very difficult to handle the situation if one is seeing several sibs for prediction simultaneously or sequentially and does not know what, if anything, they have shared in terms of results. Even more difficult is when a person refuses disclosure of a result to their spouse or partner, especially when children have already been born or a pregnancy is in progress. Our centre's general policy has been the same in HD as for genetic disorders in general: to encourage the individual to disclose information to others directly affected by it, but not to breach confidence apart from the exceptional situations when this has failed and there is a clear necessity for the individual requesting information to have it. Often the problem can be got around in an indirect manner; thus we have frequently been able to confirm to a person requesting HD prediction or to the clinician involved that the HD mutation is indeed present in the family but without stating which particular member of the family the statement refers to.

The final category of third party requests is from outside the family. Insurance requests have already been mentioned, and are difficult, because they usually come with a signed consent for release of 'medical information'. Whether predictive genetic information is indeed medical, and whether a general consent is adequate can both be questioned, as discussed in Chapters 1 and 4. My own policy is never to send the information direct, but to reply to the company that I consider the request for genetic information inappropriate, that I have sent the written result to the individual, who is welcome to give it to the company if they wish.

Requests from employers have so far been few, but have included a request from a construction firm for a predictive test result in a healthy person at risk who had an accident at work. Perhaps more worrying have been requests from

Table 3.3. Categories of late-onset genetic disorder

(a) *HD and other neurological degenerations*
Huntington's disease
Hereditary ataxias
Familial Alzheimer's disease
Familial prion dementias
Familial motor neurone disease
Familial amyloid neuropathy
(b) *Late-onset genetic disorders involving other systems*
Mendelian cancer syndromes
Familial adenomatous polyposis
Von Hippel Lindau disease
Multiple endocrine neoplasia
Adult polycystic kidney disease
Myotonic dystrophy
Haemochromatosis (autosomal recessive)
(c) *Common, largely non-genetic disorders, with Mendelian subset*
Breast cancer
Colorectal cancer
Alzheimer's disease
(d) *Common disorders with genetic susceptibility*
Diabetes mellitus
Schizophrenia
Manic-depressive illness
Coronary heart disease
Hypertension

prison medical officers for testing of prisoners. In one case of serious violent crime, the request was made even though the individual, at 50% risk for HD, had already been examined by both a neurologist and a psychiatrist, with normal results. Enquiry 3 years later showed that this remained the situation. Again this illustrates the tendency for professionals and others to consider anyone possessing the HD gene as already affected by the disorder. If such requests were to be channelled solely through a laboratory, serious misuse and inappropriate testing could well occur.

HUNTINGTON'S DISEASE AND OTHER LATE-ONSET GENETIC DISORDERS: SIMILARITIES AND DIFFERENCES

The issues and problems associated with presymptomatic testing for HD have been described at some length because I consider that they are all of relevance to other genetic disorders. If we consider now what these other disorders are, they can be divided into four main groups, summarized in *Table 3.3*.

Other late-onset neurological degenerations are in most respects similar to HD in the issues raised; indeed some are identical. Although most are rare, the list is growing, and there are special aspects to note. In some instances, for example, familial Alzheimers disease [21], familial Creutzfeldt-Jakob disease [22] and familial motor neurone disease [23], inherited cases form a small subset of a more frequent disorder where most cases are likely to be non-genetic in nature. This means that until a mutation is detected, relatives of an isolated case may have no idea that they are at genetic risk, unlike HD, where this is implied by the clinical diagnosis. This apart, late-onset neurogenetic disorders such as listed in *Table 3.3a* can appropriately be handled using protocols for presymptomatic testing comparable to those in service use for HD, and this approach is being widely adopted in most European countries [24, 25]. It is to be hoped that data collection and evaluation will also be undertaken in the same co-ordinated way as has proved so valuable in HD.

The next group to be considered are late-onset Mendelian disorders involving systems other than the central nervous system. This group is highly variable and even a category such as familial cancers contains widely differing clinical conditions. *Table 3.3b* lists some of them, but the list of those disorders where genetic testing is possible is growing rapidly.

Among other differences from HD and other neurological disorders, two important features in the case of familial cancers are possibility of therapy (of variable effectiveness), and ranges of onset often extending into childhood. Thus, in the case of familial adenomatous polyposis most individuals at risk (at least 90%) do wish to have genetic testing, since this may allow early detection of treatable tumours (especially in polyposis and colorectal cancer), while for those who prove genetically normal, there is the obvious benefit of avoiding the need for unpleasant and invasive tests on a lifelong basis [26]. In the case of adult polycystic kidney disease, by contrast, there has so far been little uptake of DNA-based presymptomatic testing; whether this reflects the use of ultrasound in early diagnosis, the limited benefit of early therapy or other factors is not clear.

A third category (*Table 3.3c*), overlapping with the previous two, but deserving separate consideration, is the group of common disorders in which most cases are isolated and possibly largely non-genetic in causation, but where an important genetic subset exists, usually following dominant inheritance. Colorectal [27] and breast cancer [28] currently provide the best documented examples, with around 5% of cases due to specific genes giving high risks to relatives, and characterized by a tendency to early-onset and multiple tumours, as well as by familial aggregation. Alzheimer's disease could be considered in this group also, but here the proportion of clearly Mendelian families is extremely small.

The particular challenge provided by this group for genetic testing is how to distinguish the Mendelian subset, where genetic tests may be highly applicable and of great importance, from the great majority of cases where this is not the case. This is particularly highlighted by the situation for breast cancer, where uncertainties surrounding the value of early diagnosis and treatment mean that in one series only around 50% of those in a known high risk group wished for genetic testing [29].

The final group of late-onset genetic disorders to be considered (*Table 3.3d*) is where there is no clear Mendelian pattern, but where it is evident from twin and adoption studies, or other genetic evidence, that a considerable genetic component is involved. *Table 3.3d* lists some examples, though they may not all be 'late-onset' in nature. It is likely that in some of these, specific genes of major effect will be identified (e.g. some types of diabetes) [30] while in others, this may not be the case (possibly schizophrenia) [31]. The common feature in this group is that possession of a specific genotype does not necessarily imply that one will develop the disorder, as in HD. Risks may be increased or diminished but the situation is not definitive. It is likely that some common disorders (e.g. coronary heart disease) will prove to fit both into this category and the previous one, with a few genes involved being of major effect (e.g. familial hypercholesterolaemia), while others have a lesser and more quantitative effect.

Quite how predictive genetic testing will prove useful in this potentially very large group is far from clear. It may be that particular therapeutic approaches could be targeted to a specific genotype, but this remains to be proved. Likewise, the reproductive reasons that rank highly in requesting testing for HD and comparable disorders are not likely to be a significant factor here. In the absence of clear information as to the likelihood of benefit (or harm), it would seem wise that if predictive testing in this category should be used at all, it should be carried out as part of a thorough research evaluation before being brought into service use. Chapter 7 enlarges on some of the issues involved in genetic susceptibility testing.

PLANNING A PRESYMPTOMATIC TESTING SERVICE FOR LATE-ONSET GENETIC DISORDERS

The initial experience with HD presymptomatic testing, from which we have all learnt so much, was based on research evaluations done in experienced centres, often with extra clinical and psychologist staff to give a substantial team involved. Now that this initial phase is largely over, testing in HD has become part of regular service provision, at least in the UK, and considerable adaptation has been required to deliver satisfactory testing without the additional research staffing. This has on the whole been beneficial; centres have had to rethink their plans, remove unnecessary elements, and consider seriously which aspects are of importance to those being tested, rather than just to those who are running the programme.

The upshot has been the evolution of a 'slimmed down' testing procedure that

Table 3.4. Service structure for HD presymptomatic testing

Interview 1 (1 hour)
Discussion of HD and its inheritance
Family details
Reasons for requesting testing
Nature and limitations of testing
Present and future ways of coping with HD and test results
Clinical assessment
Blood sample 1
Interview 2 (45 min)
Responding to queries remaining or arising from interview 1
Explanation of possible results
Review professional and social support
Practical arrangements for result giving
Consent form signed
Blood sample 2
Result giving
Follow-up
1 week – telephone contact
1 month – home visit (counsellor)
3 months – telephone contact (visit if needed)
1 year – clinic visit (gene carriers)

can be performed within the existing staffing and budget of a comprehensive medical genetics service, though it is demanding and requires careful planning. *Table 3.4* summarizes our own Cardiff service protocol, based on that recommended by the UK Huntington's Disease Prediction Consortium [31], a group that has been formed by all UK centres involved in this activity (around 20 in total).

The staff involved are one clinical geneticist and one co-worker (normally a genetics nurse specialist), both having other general genetic counselling responsibilities as well as those for HD. It should be noted that the first interview assumes that general aspects have already been resolved. Most requests for genetic testing arise in the context of a more general need for information about HD, and it seems unwise to focus the agenda specifically on testing until it is quite clear that this is really what the individual wishes.

We have found that two interviews (each linked to taking of a blood sample), are extremely valuable in allowing time for thought, understanding of the new information given at the first interview, and sometimes reassessment of the person's wishes. Not only would it be close to impossible to compress all the information into a single interview, but this would remove what is perhaps the most important aspect of the two stage process – to give time for reflection.

We have found the actual result-giving interview usually to be brief; whether it is good or bad news that has been imparted, those given the result and their partner or other person accompanying them rarely wish to stay around, but to return home so that they can express their feelings fully.

The subsequent interviews after result giving, are likewise less extensive in this service model than it might appear. The initial contact after 1 week is by telephone, that at 1 month usually a home visit by the genetic counsellor. We only schedule clinic visits if there seems to be a specific need that cannot be met by a home visit. Perhaps the most valuable role that a clinic visit can play is to allow regular clinical assessments for those shown to have the mutation and where the question now often becomes – "is there any sign of HD beginning?"

To what extent can this service structure be used for late-onset genetic disorders generally? I would suggest that to a large extent it can, that it provides a suitable foundation on which disease-specific information can be based, and that a simple two-step protocol should allow most of the necessary information-giving, exploration of attitudes and provision of basic support, regardless of the specific nature of the disease.

Of course it may be suggested that it is quite impractical for this amount of time to be given to the discussion of presymptomatic testing by a busy hospital clinician or family doctor during their regular clinic sessions. This is indeed so, but it can be argued in response that this is not an activity that anyone should attempt to undertake unless they have the time or training to do it thoroughly. Certainly, genetic centres see people where lasting harm has resulted from time not being given to careful explanation and answering of questions.

As predictive genetic testing becomes available for an increasing range of late-onset disorders and involves progressively more clinical specialties, the issues discussed in this chapter will become ones that all clinicians need to be aware of, as well as those providing laboratory services. I would suggest that the patterns of practice that have served us well for HD, widely accepted as providing some of the most difficult situations in genetic counselling and testing, will also provide a valuable foundation for presymptomatic testing services for the majority of other serious, late-onset genetic disorders.

REFERENCES

1. Harper PS (1996) (ed.) *Huntington's Disease*, 2nd Edn. W.B. Saunders, London.
2. Huntington's Disease Collaborative Research Group (1993) A novel gene containing a trinucleotide repeat that is expanded and unstable on Huntington's disease chromosomes. *Cell*, **72**: 971–983.
3. Gusella JF, Wexler NS, Conneally PM *et al.* (1983) A polymorphic DNA marker genetically linked to Huntington's disease. *Nature*, **306**: 234–238.
4. Bloch M, Adam S, Wiggins S, Huggins M, Hayden MR (1992) Predictive testing for Huntington's disease in Canada: the experience of those receiving an increased risk. *Am. J. Hum. Genet.* **42**: 499–507.

5. Tyler A, Ball D, Craufurd D (1992a) Presymptomatic testing for Huntington's disease in the UK. *Br. Med. J.* **304**: 1592–1596.
6. World Federation of Neurology Research Group on Huntington's Disease (1993) Presymptomatic testing for Huntington's disease: a world-wide survey. *J. Med. Genet.* **30**: 1020–1022.
7. Kremer B, Goldberg P, Andrew SE *et al.* (1994) Worldwide study of the Huntington's disease mutation. *N. Engl. J. Med.* **330**: 1401–1406.
8. Tyler A, Morris M, Lazarou L, Meredith L, Myring J, Harper PS (1992) Presymptomatic testing for Huntington's disease in Wales 1987–90. *Br. J. Psychiat.* **161**: 481–488.
9. Morris M, Tyler A, Harper PS (1989) Problems in genetic prediction for Huntington's disease. *Lancet*, **9**: 601–603.
10. European Community Huntington's Disease Collaborative Study Group (1993) Ethical and social issues in presymptomatic testing for Huntington's disease: a European Community Collaborative Study. *J. Med. Genet.* **30**: 1028–1035.
11. Scourfield J, Soldan J, Gray J, Houlihan G, Harper PS (1997) Huntington's disease: psychiatric practice in molecular genetic prediction and diagnosis. *Br. J. Psychia.* **170**: 146–149.
12. Morris M, Tyler A, Harper PS (1988) Adoption and genetic prediction for Huntington's disease. *Lancet*, **i**: 1069–1070.
13. Maat-Kievit JA, Vegter van der Vlis M, Zoeteweij ME *et al.* (1995) Predictive testing of 25% at risk persons of Huntington's disease. In: *16th International Meeting of the World Federation of Neurology Research Group on Huntington's disease*, p. 30 (abstract).
14. Harper PS (1993) Research samples from families with genetic diseases: a proposed code of conduct. *Br. Med. J.* **306**: 1391–1394.
15. Harper PS, Soldan J, Tyler A (1996) Predictive tests in Huntington's disease. In: *Huntington's Disease* (ed. P.S. Harper) pp. 395–424. W.B. Saunders, London.
16. Decruyenaere M, Evers-Kiebooms G, Boogaerts A *et al.* (1995) Predictive testing for Huntington's disease: risk perception, reasons for testing and psychological profile of test applicants. *J. Genet. Couns.* **6**: 1–13.
17. Huggins M, Bloch M, Kanani S *et al.* (1990) Ethical and legal dilemmas arising during predictive testing for adult-onset disease: the experience of Huntington's disease. *Am. J. Hum. Genet.* **47**: 4–12.
18. Quaid KA (1993) Presymptomatic testing for Huntington's disease in the United States. *Am. J. Hum. Genet.* **53**: 785–787.
19. Marteau T, Richards M (eds) (1996) *The Troubled Helix*. Cambridge University Press, Cambridge.
20. Codori AM, Brandt J (1994) Psychological costs and benefits of presymptomatic testing for Huntington's disease. *Am. J. Med. Genet.* **54**: 174–184.
21. Sherrington R, Rogaeu EI, Liang Y *et al.* (1995) Cloning of a gene bearing missense mutations in early-onset familial Alzheimers disease. *Nature*, **375**: 754–760.
22. Doh-ura K, Tateishi J, Kitamoto T, Sasaki H, Sakaki Y (1990) Creutzfeld-Jakob disease patients with congophilic kuru plaques have the missense variant prion protein common to Gerstmann–Straussler syndrome. *Ann. Neurol.* **27**: 121–126.
23. Rosen DR, Siddique T, Patterson D *et al.* (1993) Mutations in Cu/Zn superoxide dismutase gene are associated with familial amyotrophic lateral sclerosis. *Nature*, **362**: 59–62.

24. Tibben A, Stevens M, de Wert GMWR *et al.* (1997) Preparing for presymptomatic DNA testing for early onset Alzheimer's disease/cerebral haemorrhage and hereditary Pick Disease. *J. Med. Genet.* **34**: 63–72.
25. Lennox A, Karlinsky H, Meschiro W *et al.* (1994) Molecular genetic predictive testing for Alzheimer's disease: deliberations and preliminary recommendations. *Alzheimer's Dis. Ass. Disord.* **8**: 126–147.
26. Burn J, Chapman P, Delhanty J, Wood C, Lalloo F, Cachon-Gonzalez MB, Tsioupra K, Church W, Rhodes M, Gunn A (1991) The UK Northern Region genetic register for familial adenomatous polyposis coli: use of age of onset, congenital hypertrophy of the retinal pigment epithelium, and DNA markers in risk calculations. *J. Med. Genet.* **28**: 289–296.
27. Bellacosa A, Genuardi M, Anti M, Viel A, Ponz de Leon M (1996) Hereditary non polyposis colorectal cancer: review of clinical, molecular genetics, and counselling aspects. *Am. J. Med. Genet.* **62**: 353–364.
28. Castilla LH, Couch FJ, Erdos MR *et al.* (1994) Mutations in the BRCA1 gene in families with early onset breast and ovarian cancer. *Nature Genetics* **8**: 387–391.
29. Lerman C, Bieseke B, Benkendorf JK, Gomez-Caminero A, Hughes C, Reed MM (1997) Controlled trial of pretest education approaches to enhance informed decision making for BRCA1 gene testing. *J. Nat. Cancer Inst.* **89**: 148–157.
30. Velho G, Froguel P, Clement K *et al.* (1992) Primary pancreatic beta-cell secretery defect caused by mutations in glucokinase gene in kindreds of maturity onset diabetes of the young. *Lancet* **340**: 444–448.
31. McGuffin P, Owen MJ, Farmer AE (1995) Genetic basis of schizophrenia. *Lancet* **346**: 678–682.
32. Craufurd D, Tyler A (1992) Predictive testing for Huntington's disease: protocol of the Huntington's Prediction Consortium. *J. Med. Genet.* **29**: 915–918.

Chapter

4

GENETIC TESTING AND INSURANCE

P.S. Harper

INTRODUCTION

Among the various social implications of new developments in genetics, fear of discrimination in the field of insurance has given more concern than almost any other issue. It is an obvious practical issue, affecting most individuals at some point in their life. In the UK, life insurance is essential to cover a loan in buying a house, and is thus a matter of great importance to young adults. By contrast in the USA, most of the debate has resolved around health insurance, since there is no national health service and since health insurance is increasingly linked to employment. The content of this chapter may thus seem rather uneven to the American reader, and those wishing to redress the balance should consult the numerous articles that make genetic testing in relation to health insurance the main focus.

The topic has been one that I have found particularly frustrating, since from the outset, I have felt strongly that it should not, and need not be a problem at all. From the time of my first involvement over 5 years ago, I have believed that a practical solution was possible regarding life insurance, that would involve insurance companies in minimal risk or loss, and that would allow the privacy of pre-existing genetic test results to be maintained. Such a practical solution, restricted to life insurance sums of a reasonable amount to allow house purchase and normal cover of family responsibilities, seems to me an essential factor in maintaining an equitable society that does not create impossible burdens for those at risk of genetic disorders.

Sadly, my involvement in a series of debates, arguments and abortive discussions with the insurance industry has led me to conclude that the industry, with one or two exceptions, basically does not wish to consider the matter seriously, nor to move from its initial entrenched position that it must have access to all test results and the right to use them, and that genetic test information is no different from 'other medical information'. I suspect that these attitudes have been strongly influenced by views in the USA, largely based on health insurance problems, and that the UK insurance industry has not been prepared to think through the issues beyond the point of intense reluctance to accept any form of regulation or compromise.

The first of the two papers given here, written in 1993, still seems to give a valid account of the situation existing today, since little has happened subsequently to alter the situation. The second paper, written in 1996, explores the critical issue of adverse selection further. I had originally hoped that collaboration with experts inside the insurance industry would make it possible to cost accurately the overall risks and benefits of ignoring genetic test results, but in the absence of this, I went ahead in the hope that experts might be provoked into producing a more detailed analysis. In the event, the first actuarial estimates of the effects of genetic testing were presented alongside this paper at a joint conference of geneticists and actuaries, as described at the end of this chapter. Both the 1993 and 1996 papers are given in full here, since they overlap very little, and the main issues have not significantly changed.

INSURANCE AND GENETIC TESTING[a]

Genetic testing in relation to insurance was first raised by R.A. Fisher in 1935, in an address to the International Congress of Life Assurance Medicine, but it is only now that molecular techniques have made such testing a reality. More and more serious disease-causing genes are being mapped and isolated, with a corresponding increase in our ability to predict or exclude them in family members at risk. Concern that genetic testing might harm both those with a family history of a genetic disorder and others first emerged in the USA [1], where many people depend on private insurance for health. In Europe (e.g. The Netherlands) the issue has also been recognized [2,3] and legislation in the wider context of the European Community is a possibility. In the UK the topic received little public attention until it was raised after the 1991 International Workshop on Human Gene Mapping, held in London [2], since when it has had widespread media coverage and professional discussion [4], even though little use has yet been made of genetic testing for insurance purposes. The insurance companies, those being or likely to be tested, and professionals in medical genetics see the issue from different angles, and all three groups have valid concerns (*Table 4.1*) that need to be resolved if confrontation is to be avoided.

The term 'genetic test' is often used loosely to include not only tests based on DNA analysis of genes but also cytogenetic tests for chromosome abnorma-

[a]Originally published by *Lancet* (1993) Vol. 341, pp. 224–227, and reproduced with their kind permission.

Table 4.1. Principal concerns of insurers, applicants and professionals

Insurance companies
Adverse selection
Competition
Avoidance of unnecessary discrimination
Avoidance of adverse public opinion and legislation
Those at risk
Pressure to be tested to obtain basic life and health insurance
Confidentiality of high-risk result
Discrimination and stigmatization after high-risk result
Wish to obtain insurance after low-risk result
Professional concerns
Testing of individuals who would otherwise prefer not
Deterrence of those who would otherwise wish to be tested
Testing without adequate counselling
Testing minors
Stigmatization of those with genetic diseases
Pressure for termination of affected pregnancies

lities and tests for specific gene products (see Chapter 1). For the purpose of discussion here the term will be restricted to DNA analysis. I must also emphasize that it is the predictive use of such tests in the healthy individual, who may be at risk genetically, that is the issue – not their use in the investigation of a person already ill.

Insurance is based on the complementary principles of solidarity and equity in the face of uncertain risks. Solidarity implies the sharing by the population, as a whole or in broad groups, of the responsibility and the benefits in terms of costs, while equity means that the contribution of an individual should be roughly in line with his or her known level of risk. Life insurance (assurance) and health insurance both have medical implications relevant to genetic testing. In countries with a comprehensive national health-care system health insurance is currently less important than in countries such as the USA, where private health insurance, increasingly employer-based, is the principal means of paying for health care. The European situation could change radically if national schemes were to be replaced or downgraded. Life insurance is a more uniform need in developed countries, being closely linked to family responsibilities and, in the UK, to house purchase. Life insurance is principally relevant to disabling or fatal conditions of adult onset such as Huntington's disease and other progressive neurological diseases (though premiums may be raised for conditions whose effects are mild or uncertain); health insurance, by contrast is important for chronic and disabling genetic disorders in all age groups, especially those of childhood. There are other forms of insurance, more recently developed, that may involve few individuals but are of considerable importance to the insurance industry, and some of these are discussed below.

The insurer's viewpoint

UK life insurance companies have no policy at present of actively asking for genetic tests but the results of any tests that have been done must be made available and may be acted on in assessing risks. In the words of the Association of British Insurers' 1992 memorandum:

> From the point of view of insurers, genetic diseases can be divided into two main groups. The first is the known genetic disease such as Huntington's chorea, Cystic fibrosis or Duchenne's muscular dystrophy, for which specific tests are already in use. In these cases the insurance industry already has experience of individuals who have a genetic test because of medical history and insurers treat the results of such tests in exactly the same way as the results of any other medical test. The UK insurance industry does not intend to ask proposers for life insurance to undergo screening for genetic information within the foreseeable future, but where individuals have had a specific test as part of their medical assessment these tests will fall into the same category as other medical tests and must be declared on proposal forms.

It would thus seem that there is a clear duty expected for individuals and their doctors to disclose the results of any genetic test, though it could be questioned as to whether a test done in the context of genetic counselling should be regarded as part of a 'medical assessment'. A comparable position has been taken by insurance companies in the USA; genetic tests are considered in the same light as any other medical test [1].

Adverse selection is the foundation of all the fears that insurance companies have in relation to genetic testing. They see it as unfair for an applicant for insurance to have information that the company does not have. Adverse selection is especially feared by insurance companies in relation to large sums insured. It is also relevant to 'critical illness' insurance, where a large sum may be payable in the event of the applicant developing one or more named serious diseases. LRG Services (a Lloyds-based company) have offered a 'Babycover' policy which will pay out if a baby is born with, or develops during the first 2 years of life, a disorder such as muscular dystrophy, cystic fibrosis, Down's syndrome or spina bifida.

Commercial competitiveness is intense and the fear is that if one company does not use genetic testing information but its competitors do, that company would lose custom from those proving to be at low risk while carrying the increased burden of those with an undisclosed high risk. On the other hand, companies wish to avoid unnecessary testing, partly on grounds of medical and administrative work but also to avoid losing customers. In the UK over 95% of life insurance policies are at the standard premium rate; less than 1% of proposals are refused because the risk is deemed too high. A study by the US Office of Technology Assessment [5] suggests that companies would in general neither request nor use information from genetic tests at present, but they would wish to have such information if it were already available. Any widespread genetic testing that excluded a large proportion of the population would not be in the insurance companies' interest.

The applicant's concerns

The concerns commonly expressed by those at risk for a genetic disorder are summarized in *Table 4.1* but this is not the group of greatest interest to insurance companies. They are mainly concerned with people without a known family risk who might obtain knowledge of such a risk as a result of availability of genetic testing. It is difficult to assess how seriously pressure to undergo genetic testing to obtain insurance is affecting decisions now or what may happen in the future. A UK study of reasons underlying the decision to have a predictive test for Huntington's disease (HD) found insurance ranking lower than other personal and reproductive reasons [6]. The stated policy that insurance companies do not intend to seek testing also suggests that such pressure will remain slight. However, relatives of patients with a serious disorder such as HD usually find life insurance impossible to obtain or the premium greatly increased; and in a situation where a person cannot provide for their family's future or purchase a house, pressure to be tested in the hope of a favourable result must be great.

A worrying issue now is the disclosure of results of genetic tests done for reasons unrelated to insurance, the basis for almost all the tests so far undertaken for HD and other disorders. Not only do individuals at risk fear that such disclosures, by themselves or by their doctor, would make any future insurance even more difficult, but also there is a concern about confidentiality, even though insurance companies consider themselves to have a good record on this point. The danger is that the information might reach employers (a particular concern in the USA, where employers are increasingly responsible for arranging health insurance). There is also the risk of adverse consequences from a finding of a harmless carrier state in recessive disorders such as cystic fibrosis or the haemoglobinopathies or from uncertainty about the clinical significance of findings (different mutations for cystic fibrosis or 'premutations' in myotonic dystrophy and fragile X mental retardation are examples). Confusion on this point is clearly widespread since in the American survey [5] just over half of those companies asked considered the carrier state for such conditions as cystic fibrosis or Tay–Sachs disease to be a 'pre-existing condition'.

The concerns of professionals in medical genetics

Much of what we know about people's attitudes to genetic testing and the effects that it has had on them and their families has come from research into HD. Predictive testing, originally by linked genetic markers, has been available since 1986, this being accompanied by careful preparation and counselling. Experience in America and Europe has shown that while this testing has a major impact on those involved, there have been few major adverse effects (see Chapter 3). Only a small minority of those at risk have in fact requested testing and many of those inquiring about the possibility have decided not to pursue the idea. Although the main reasons for testing have been to do with resolving uncertainty or family planning [6], the possible insurance-related use of such tests has generated considerable debate. Unpressured consent is a cornerstone of this form of testing so misgivings about the insurance-related use of testing are not surprising; even if a test is not requested by an insurance company, the use of results to alter risk calculations could become a pressure for individuals to be tested who might not otherwise have wished for this and who in some cases may be vulnerable to a high-risk result. If disclosure of results were mandatory, the converse might well also be seen; individuals who wish for testing might not go ahead for fear that the result would be disclosed by their doctors to insurance companies or that they might themselves later be penalized for non-disclosure.

Testing of minors is especially difficult (see Chapter 2), notably for late-onset, currently untreatable disorders such as HD. Although unlikely to be relevant to life insurance, this could become a major issue in health insurance; for example, testing of children known to be at risk from serious disorders such as cystic fibrosis, could be made a prerequisite for health insurance acceptance.

The above concerns could be much more widespread and more serious in their consequences in the context of population screening. In population programmes it is easy for counselling to be neglected, so that people get results they had not fully consented to, or at least not fully understood the implications of; such individuals might then find themselves disqualified from insurance. Also with genetic tests done for research, the finding of an abnormality may be unforeseen and its implications often only partly understood.

General issues

What general points might meet with the agreement of all involved?
Genetic disease is currently not increasing, a situation contrasting with HIV infection where insurance companies have faced large payments in an age group which previously was relatively healthy. Indeed the reverse is true, especially for those serious disorders amenable to genetic testing. Individuals have used the results to avoid the birth of affected children, through a combination of carrier detection and prenatal diagnosis, thus decreasing the likely future burden on insurance companies.

How great is the problem of adverse selection? As I have shown, there is a clear difference between the perceptions of the insurers and those of families and health-care professionals. The immediate concern is the relative with a known family risk of a serious genetic disease. At present, for a disorder such as HD, insurance cover is usually impossible for a person at 50% risk but such a person is already largely disbarred from insurance on grounds of family history so an undisclosed high-risk result will be of little relevance to an insurance company. Correspondingly, a low-risk result, while benefiting the person now able to be insured, has no adverse effect on the insurer. Adverse selection here would only arise if an individual declared neither family risk nor high-risk result – but this can happen already and non-disclosure of family risk would invalidate a contract with an insurance company. For less severe disorders, where those at 50% risk can obtain insurance at increased premium, the scope for adverse selection would be limited by the difference in premium and benefits for the person at 50% risk and one known to be affected; the sums involved are probably not large. Adverse selection in the context of a known family risk is thus not likely to be serious, an important conclusion since most genetic testing is in this context.

Do we wish to avoid widespread insurance discrimination on the basis of genetic tests? The insurance companies' policy is to avoid selection unless clearly necessary, and the American health insurance industry has even been criticized for waste by making excessive demands for such discriminatory tests [7]. The view of the general public is probably similar. Most professionals are also likely to be against discrimination on grounds of genotype. The editor of the journal *Nature* has taken the opposite view, arguing that "The contention that people unlucky enough to carry identifiable genetic abnormality should not be denied insurance on the same terms as other people, begs the question why people relatively free from identifiable genetic abnormality should therefore pay more than would otherwise be necessary" [8]. This stand may be tenable in relation to rare Mendelian disorders but certainly not when testing for susceptibility to a range of common disorders becomes feasible.

Health insurance

In Europe most people are not dependent on private health insurance but the opposite is true in the USA, and genetic testing in relation to health insurance has prompted much debate [9,10]. The Office of Technology Assessment's important report [5] provides valuable information on this, while its earlier report on cystic fibrosis screening [11] covers the wider issues of genetic discrimination and stigmatization. A disturbing aspect of the US experience is misconception about the results; heterozygote carriers may be regarded as having disease [5], as happened previously with sickle cell testing. Confusion is likely to increase as more and more disorders, often rare and unfamiliar, come into the picture. The consequences of denial of health insurance may be immense where there is no comprehensive national system of health care and could even extend to loss of employment. This largely US concern could spread to Europe especially when governments attempt to shift the balance of health care from the public to the private sector.

The way forward

Genetics professionals are upset that a situation that many of them had foreseen has been largely ignored and that opportunities for constructive dialogue have been lost. Is it too late? perhaps not, but the 'window of opportunity' is closing. The following suggestions are put forward until a more comprehensive framework can be constructed.

(i) For any large or unusual life insurance application, companies should have the right to ask about or have access to results of relevant genetic tests. The ceiling could be debated, but in The Netherlands 200 000 guilders (£50 000) has been suggested [3].

(ii) For all normal life insurance applications, up to an agreed level, companies should neither request nor require disclosure of genetic test results. Agreement on this might remove the need for legislation, which has been threatened in Europe [2] and the USA [10,12]. In The Netherlands there is an agreement (to be reviewed after 5 years) that ordinary life insurance cover under 200 000 guilders should not require such disclosure.

(iii) Where health insurance is an integral part of health-care provision the non-disclosure principle as for life insurance is essential.

(iv) Doctors should not feel obliged to disclose to insurance companies the results of existing genetic tests done on healthy individuals with a family history of a genetic disorder.

(v) Population genetic screening should be avoided if it is likely to have insurance implications, and batteries of genetic tests should not be offered to those not at specific risk.

Detailed discussion between the parties involved should now take place. To allow an issue of such practical and ethical importance to drift into unconsidered and piecemeal applications is unacceptable.

In the 1993 paper given above, 'adverse selection' was discussed briefly, but in the following years it developed a position of central importance in the debate, with insurance industry representatives claiming that it was of such magnitude that it could destroy life insurance as presently organized, or at least cause large rises in premiums if genetic test results were not disclosed. I tried to develop the subject in the following paper given to a joint meeting of the Royal Society and the Institute of Actuaries in late 1996.

GENETIC TESTING, LIFE INSURANCE AND ADVERSE SELECTION[a]

The use of genetic test information in relation to life insurance remains a controversial topic. Although the insurance industry has not so far actively

[a]Reproduced from *Human Genetics – Uncertainties and the Financial Implications Ahead* (1997), Transactions of the Royal Society (B), Vol. 352, by kind permission of the editor.

requested genetic tests, it has been reluctant to move from its established position of considering genetic tests as no different from other medical information; that it needs to know all existing results and, where appropriate, to act on them. The industry has also been reluctant to have serious discussion on the issue with professionals in medical genetics, despite long-standing calls for this and has maintained that the freedom to underwrite is an essential feature of life insurance provision [12]. Medical geneticists and those representing families with genetic disorders, by contrast, have highlighted the danger to those at risk for genetic disorders of being deterred from potentially beneficial tests for fear of insurance penalization and possible wider disclosure of sensitive information [13].

The UK Parliamentary Select Committee on Human Genetics was critical of the insurance industry for its refusal to face up to the important issues involved. Their report, issued in July 1995 [14], suggested a 1-year period in which to resolve the situation, after which legislation might be considered. While this has not so far been acted upon through the government's response to the report [15], insurance industry bulletins have stated their intention to fulfil the report's recommendation [16].

The principal concern of the insurance industry relates to 'adverse selection', the term applied to financial losses incurred as a consequence of an applicant having information to which the insurance company does not have access. This is a genuine concern, but there are no valid estimates as to its likely extent, either at present or in the future, in genetic testing.

As a starting point to obtaining an estimate of adverse selection, I outline here the main categories of genetic disease and the extent to which adverse selection is likely to be important in them. Hopefully, these general data can be

Table 4.2. Life insurance and the main types of genetic disorder

Category of genetic disorder	Relevance to life insurance
Autosomal dominant	Important sub-group with late-onset and progressive course. Those with early onset of little relevance
Autosomal recessive	Little relevance. Genetic risks largely confined to sibs. Often early onset. Numerous healthy carriers
X-linked	Risks mainly to male relatives; serious disorders usually have early onset
Chromosomal abnormalities	Usually early onset, not progressive, carriers normally healthy
'Multifactorial' disorders	Common; genetic testing of uncertain significance at present, but likely to be important in future

converted by others into more detailed costs, based on actuarial information. It should be noted specifically that this discussion relates to life insurance, not to health insurance, critical illness cover or other special schemes, and to clearly genetic (i.e. 'Mendelian') disorders, not to susceptibility testing for 'multifactorial' conditions. The insurance industry has itself stressed that it considers tests in this latter group to be insufficiently validated to apply in life insurance at present [17], though its attitude on this may now be changing.

Categories of genetic disorder

Table 4.2 lists the main groups of genetic disorder used by clinical geneticists and others. Leaving aside the 'multifactorial' category, only one sub-group – dominantly inherited disorders of late onset – is likely to be significant in terms of adverse selection. It is thus important to explain in some detail why the other groups, accounting for most serious genetic disease, are not relevant.

(i) *Autosomal recessive disorders.* These include many serious childhood diseases giving early mortality and major morbidity. It is this early onset that minimizes their relevance to life insurance – healthy sibs reaching adult life without clinical features are unlikely to become affected later, at least to any degree affecting mortality. Also, it is only sibs, not offspring or more distant relatives that are at significant risk; most genetic tests in healthy relatives are done in relation to reproductive risks to establish whether an individual is a heterozygous carrier, which generally carries no health implications but which could easily be misinterpreted if the information had to be declared for insurance purposes.

(ii) *X-linked disorders.* Numerically much less frequent than autosomal conditions, this group nevertheless contains some serious diseases. However, for life insurance purposes, the implications are again few. Most fatal disorders are early in onset (e.g. Duchenne muscular dystrophy), while even those giving serious problems in adult life (e.g. haemophilias A and B, Becker muscular dystrophy) are usually obvious by the end of adolescence. Others, such as fragile X mental retardation syndrome are not life threatening. As with autosomal recessive disorders, most genetic tests on healthy adults are done in relation to carrier status, with no significant implications for health of the individual. I have been able to find no X-linked disorder apart from a few extremely rare conditions, that would give a likelihood of serious adverse selection in relation to genetic tests.

(iii) *Chromosomal disorders.* Chromosome analysis remains the most frequent form of genetic test, but is of little relevance to life insurance. Disorders of autosomes with a visible defect are generally severe and obvious at birth or in infancy, whereas those of sex chromosomes do not significantly affect later health. As with the previous groups, testing of healthy adults is done for reproductive rather than health reasons. Adverse selection is thus not an issue in this group.

Table 4.3. Autosomal dominant disorders relevant to life insurance

Disorder	Comments
Huntington's disease	Healthy individuals only tested in context of family history
Hereditary ataxias (late onset)	As for HD
Other rare, late-onset CNS degenerations	All very rare. Combined frequency unlikely to exceed that of HD
Myotonic dystrophy	Mutation variable. Many tested in later life will have insignificant disease
Adult polycystic kidney disease	Little current demand for genetic testing. Early diagnosis by ultrasound
Familial colon cancer (polyposis and non-polypotic)	Mortality much reduced by early detection
Familial breast cancer (BRCA1 and 2)	Important implications if widespread screening introduced
Other rare familial cancers	All very rare; some treatable
Marfan syndrome	Usually evident clinically; genetic testing secondary
Hypertrophic cardiomyopathy	Interpretation of tests and risks uncertain
Familial hypercholesterolaemia	Common, but usually detected by serum cholesterol, not genetic tests

(iv) *Autosomal dominant disorders.* This is the only group where there is potential for serious adverse selection and this is the case only where the disorder is late in onset but progressive and fatal in nature. Early-onset dominantly inherited disorders will not give significant risks for apparently healthy relatives and many will be isolated cases resulting from new mutation.

Some of the most important members of the late onset, progressive group are listed in *Table 4.3*. It includes progressive neurodegenerative diseases such as HD and allied disorders, cardiovascular disorders such as Marfan syndrome with a risk of sudden death in later life, and several forms of familial cancer. Careful study of McKusick's *Mendelian Inheritance in Man* [18] shows that the number of disorders in this group, other than those that are exceedingly rare, is limited. Those listed comprise the principal ones, though it is possible that in a particular geographical area, some others might be sufficiently common to join the list.

Several of the disorders listed are, currently at least, less of a problem in relation to life insurance than might be thought likely at first sight. Thus, Marfan syndrome can almost always be diagnosed or excluded in adult life by careful clinical assessment, while adult polycystic kidney disease can similarly be recognized presymptomatically by ultrasound. Genetic tests are currently little used in either. Variability and heterogeneity may also limit interpretation of genetic testing, as in hypertrophic cardiomyopathy where ultrasound examination is again a more helpful guide to future clinical abnormality than are genetic tests.

A feature common to all the disorders in *Table 4.3* is that the great majority of cases occur within the context of a family history of the condition; presymptomatic genetic testing is performed almost exclusively in the presence of such a history. Since most life insurance proposal forms request information about parents, and since proposals are currently usually declined or severely loaded in the presence of such a family history, it is clear that the scope for adverse selection is extremely limited.

Treatment and early diagnosis of genetic disorders

One of the main concerns of medical genetics professionals is that worries about insurance may deter those at risk of a treatable genetic disorder from early genetic tests that could improve their prognosis, a situation that would also be against the interests of the insurance industry. Data on the outcome of such early detected disease are few, even where treatment exists, but *Table 4.4* indicates in selected examples how availability of treatment might affect the situation. The familial cancers provide the most promising example, notably adenomatous polyposis coli; evidence for benefit of early diagnosis of familial breast cancer is at present less secure.

The insurance data in *Table 4.4* are taken from the industry's own handbook [19]. The data are approximate and more detailed actuarial data may be available, or could be collected, for these examples and for other relevant diseases. The insurance industry should itself be able to estimate the financial consequences corresponding to the data.

Table 4.4 lists other factors necessary to estimate accurately the potential for adverse selection. Clearly this will be affected by the uptake of genetic testing and the frequency of the disorder; the estimates given are approximate and based mainly on the author's own centre and on the pooled data available through the UK prediction consortium for HD. It also should be remembered that the majority of tests will be normal, since age and other factors will mean that the true risk for many being tested will be considerably less than 50%. This will especially be so if non-genetic investigations (e.g. ultrasound, colonoscopy) have already detected many of the asymptomatic gene carriers. *Table 4.4* also shows that the current cautious approach of the industry to insuring healthy relatives means that the difference between a person with an abnormal test result and one with no (or an unknown) result is often small.

Diagnostic genetic testing

The life insurance issues discussed here have all presumed that the individual concerned is clinically healthy. Since specific tests for genetic mutations have

Table 4.4. Genetics and life insurance: illustrative examples

Disorder	Huntington's disease	Familial adenomatous polyposis	Adult polycystic kidney disease
Early treatment impact on mortality	Nil (at present)	Major	Moderate
Frequency (approx.)	1 in 10 000	1 in 10 000	1 in 1000
Current insurance penalty for healthy young adult at risk because of family history	Not insurable under 21 years; extra 7 per million 21–35 years [24]	Extra 5 per million if under age 35 and colonoscopy normal; probably uninsurable without colonoscopy [24]	Extra 5 per million at age 25 after normal ultrasound [24]
Current insurance penalty for clinically affected person	Not insurable [24]	Not insurable (without surgery)	Not insurable when renal function affected [24]
Mortality without early detection	Close to 100%	40% (10-year survival)	Mean survival 35 years at age 25 [20]
Mortality reduction (with early detection)	No change	93% (10-year survival)	Significantly improved
Current uptake (% of those at risk requesting genetic testing (approx.))	15–20%	90%	Low
No. of tests per year in UK (approx.)	500 (UK HD prediction consortium data)	400	100 (maximum)
Proportion of tests normal (approx.)	60% (consortium data)	>50%	>50%
No. of abnormal test results per year in UK (approx.)	200	160	40

become available, these are increasingly used *diagnostically*, in symptomatic patients who have or are likely to have a particular genetic disorder, but where its precise nature may be in doubt. In this situation, genetic testing is not directly relevant to life insurance, in contrast to *predictive* or *presymptomatic* testing, where the person is healthy but at risk for serious disease in later life. It is important to distinguish the two types of genetic testing since the implications are quite different, even though the technology may be identical.

The information given above makes it clear that the great majority of genetic disorders are of little or no relevance to the life insurance industry. Even in the small number of diseases that are of importance, genetic testing is relevant in only a small proportion. Further, in some cases, the possibility of early diagnosis or treatment could directly benefit the insurance industry by reducing mortality, while normal results will generate new potential customers. Even in those very few remaining situations where adverse selection is a real possibility, genetic testing will usually only be done in the context of a clear family history (usually an affected parent). This automatically minimizes the likelihood of unsuspected adverse selection.

On the other hand, should the life insurance industry persist in its current position of requiring all genetic test results to be declared, it will have to construct a detailed and complex system for assessing the very large and rapidly increasing amount of data relating to tests carried out primarily for reproductive reasons, where the implications to health of the individual tested are either minimal or absent. Deciding whether such information might be significant would require a large amount of specialist knowledge (several thousand Mendelian genetic disorders are documented, many extremely rare [18]) and would be likely to generate confusion and misinterpretation. Even in the few disorders where this information was potentially important (e.g. those in *Table 4.3*), interpretation could be extremely difficult. Thus a 'minimal' mutation for myotonic dystrophy might have insignificant health implications, but definition of this would be extremely complex. Increasing data on correlation of phenotype with type of mutation (e.g. in cystic fibrosis) would also have to be taken into account and could be controversial.

Thus, while I have not attempted to give an accurate estimate of potential adverse selection here, it can be seen that the scope for it is extremely limited; most adverse selection that might occur could be avoided by restrictions on large or 'critical illness' policies and by other simple and uncontroversial measures. The situation is of course entirely different to that for health insurance, though it is likely that the attitude of life insurance bodies in the UK has been influenced by this, particularly by the American situation.

Genetic testing in medical practice is at present almost entirely confined to tests for monogenic disorders, an important and numerically significant group when considered as a whole, even though few should be relevant to life insurance, as indicated above. Susceptibility testing for multifactorial disorders is currently too uncertain in terms of accurate risk prediction to be used in service, though rapid advances in identifying the genetic component of common disorders is likely to change this and lends urgency to resolving the more clear cut issues involving monogenic disorders. Genetic testing is likewise largely confined at present to those with a clear family history of a specific disorder. Should widespread population screening be undertaken this might also significantly alter the situation for adverse selection.

In first writing about this topic 5 years ago [4], I stated that there was a limited time for the industry to adopt reasonable policies on genetic testing and life insurance before applications became widespread. The House of Commons Select Committee Report likewise emphasized the urgency of finding an agreed solution. I have attempted to show here that the factor of adverse selection is not likely to be a limiting factor in reaching such an agreement; that life insurance companies could, with little loss, forgo the use of, or knowledge of, genetic test results other than in exceptional situations, and that the industry could indeed benefit from avoiding the need to assess an increasing volume of complex and largely irrelevant data.

If the industry does not itself attempt to reach a solution in consultation with those working in the field it seems likely that legislation, as already introduced in an increasing number of countries and at an advanced stage of preparation by the European Union, will result in much more restrictive use of genetic test information than is necessary or desirable for either the industry or for families with genetic diseases.

POSTSCRIPT

The topic of genetic testing and insurance remains a contentious area, with much publicity generally and considerable professional debate. Unfortunately, in the UK at least, little has changed at the practical level, despite strong pressure on the insurance industry from a variety of quarters.

The UK parliamentary report on human genetics issued in July 1995 [14] was strongly critical of the insurance industry and gave it 1 year to bring forward proposals if it wished to avoid legislation. The government's response to the report [15] specifically rejected this possibility (it is perhaps relevant that the Department of Trade and Industry was the department involved), and the year's deadline passed with no response from the industry.

An important landmark proved to be a conference in late 1996 sponsored by the Royal Society (the UK Academy of Sciences) and the Institute of Actuaries (the body representing the professionals responsible for the actuarial analyses underlying insurance), the full proceedings of which are now published [21]. My 1996 paper quoted above was presented at this meeting, and received a predictably hostile reception from an audience composed principally of insurance industry members. Much less expected, however, were the conclusions of the papers presented by actuarial experts themselves, which disagreed with statements from the insurance industry that non-disclosure of genetic test results might cause a steep rise in life insurance premiums and also concluded that the industry was probably breaking its own code of practice in collecting large amounts of irrelevant information.

Most important, was a detailed mathematical analysis by MacDonald [22] of how life insurance might be affected by adverse selection, using a variety of models based on the size of policies, the frequency of non-disclosure of important information and the different types of genetic disorder. Even considering the most extreme situations, the result was that the effects on the life insurance industry would be minimal, and that such damage as might occur would be related to large size policies, MacDonald concluded that the life insurance industry would not suffer significantly if it refrained from, or was forbidden from using genetic test information for policies other than those of large size, a conclusion very similar to that which I had suggested, without rigorous analysis.

It can be well understood that the reaction of the insurance industry audience to these conclusions, based largely on the work of their own professional experts, was even more hostile than that to my own presentation. Subsequent events have reflected considerable disarray within the industry, with some companies producing a clear statement that they no longer require disclosure of genetic test data for normal life policies, while the majority continue to hold to the view of requiring everything to be disclosed.

The long-awaited statement from the Association of British Insurers, eventually produced in early 1997, pleased no-one by its inadequacy. It suggested a 2-year moratorium on the actual use of genetic test results, solely in relation to low-level policies for house purchasing, but insisted that all information must continue to be disclosed. It is not likely that this will remove the fundamental concerns about disclosure of sensitive results, nor is it clear what might happen to the information at the end of the 2-year period.

This highly unsatisfactory and largely unnecessary chapter of events relating to life insurance has largely obscured, at least in the UK, the very real issues that genetic tests pose to other forms of insurance, notably health insurance and disability insurance. All involved agree that, in contrast to life insurance, genetic tests and consequent adverse selection could be extremely important here. The consequences of predictive tests for disorders of old age, such as Alzheimer's disease, could cause serious problems for insurance companies if those at high risk took out extensive policies covering nursing and residential care; the same problems would result if those insurance companies providing such policies were to insist on such tests being undertaken or were to exclude those at high risk.

At the other extreme of life, private health insurance could be greatly affected by genetic tests for serious childhood diseases, such as cystic fibrosis or muscular dystrophy; if prenatal tests were to have been done, or were feasible in future, pressure for termination of an affected pregnancy could be applied. These situations are already being faced in the USA, where the insurance industry is maintaining an aggressively 'free market' approach in the face of widespread criticism and threats of legislation, and denies any difference between genetic and 'other medical' information.

From a European, and particularly a UK perspective, the solution to these problems in health and disability insurance is clear – society must create and maintain responsibility for universal systems of health care and care in old age. Once such fundamental issues are reduced to the status of commodities and are left to individual provision, then pressures to individualize insurance are likely to become irresistible; with insurance companies offering cheap policies to those supposed to be genetically healthy, while penalizing or excluding those at high risk for serious genetic disease.

Fortunately, private health care in the UK remains, to a large extent, a luxury, where discrimination from genetic test results is likely to affect only a small number of individuals, in contrast to life insurance. Maintaining a high quality National Health Service is the best way to ensure that the issue of genetic testing in relation to health insurance remains a marginal one. The same situation applies to countries with universal health insurance systems, so long as they remain universal. For insurance related to disability and old age care, there are real challenges to be faced, which should be addressed now, before genetic testing becomes realistic or widespread for common disorders of later life.

As made clear in this chapter, a series of opportunities has been lost in relation to life insurance, though even at this late stage it may be possible to correct the situation. For the other, more important, forms of insurance, government, professional bodies, lay groups involved with genetic disorders and society itself must ensure that proper discussions are held and agreements reached. The insurance industry must not be allowed to dictate the agenda or to refuse to co-operate in a field which is far too important for all of us, to be left to commercial interests alone.

REFERENCES

1. Pokorski R (1989) *The Potential Role of Genetic Testing in Risk Classification. Report of Genetic Testing Committee to Medical Section of American Council of Life Insurance*, pp. 8–17. American Council of Life Insurance, New York.
2. Dutch Health Council (1989) *Genetics, Science and Society*. Dutch Health Council, The Hague.
3. Abbing HDCTR (1991) Genetic predictive testing and private insurances. *Health Policy*, **18**: 197–206.
4. Harper PS (1991) Genetic testing and insurance. *J. Roy. Coll. Phys.* **26**: 184–187.
5. US Congress Office of Technology Assessment (1992) *Genetic Tests and Health Insurance: Results of a Survey*. US Government Printing Office, Washington, DC.

6. Tyler A, Morris M, Lazarou L, Meredith L, Myring J, Harper PS (1992) Presymptomatic testing for Huntington's disease in Wales 1987–90. *Br. J. Psychiat.* **161**: 481–488.
7. Brandon RM, Podhorzer M, Pollack TH (1992) Premises without benefits: waste and inefficiency in the commercial health insurance industry. *Int. J. Hlth Serv.* **21**: 265–283.
8. Maddox J (1991) The case for the human genome. *Nature*, **352**: 11.
9. Billings PR, Kohn MA, de Cuevas M *et al.* (1992) Discrimination as a consequence of genetic testing. *Am. J. Hum. Genet.* **50**: 476–482.
10. Natowicz MR, Alper JK, Alper JS (1992) Genetic discrimination and the law. *Am. J. Hum. Genet.* **50**: 465–475.
11. US Congress Office of Technology Assessment (1992) Cystic fibrosis and DNA tests: implications of carrier screening: discrimination issues. 189–210. US Government Printing Office, Washington, DC.
12. Leigh S (1996) *The Freedom to Underwrite*. Institute of Actuaries, London.
13. Harper PS (1993) Insurance and genetic testing. *Lancet*, **341**: 224–227.
14. House of Commons Science and Technology Committee (1995) *Human Genetics: the Science and its Consequences*. HMSO, London.
15. *Government Response to House of Commons Science and Technology Committee Report* (1996). HMSO, London.
16. Association of British Insurers (1996). Public Affairs Newsletter, no. 4.
17. Association of British Insurers. Evidence to House of Commons Select Committee on Human Genetics (1995). HMSO, London.
18. McKusick VA (1992) *Mendelian Inheritance in Man*, 10th Edn. Johns Hopkins University Press, Bethesda, MD.
20. Levey AS, Pauker SG and Kassirer JP (1983) Occult intracranial aneurysms in polycystic kidney disease. *N. Engl. J. Med.* **308**: 986–994.
21. Royal Society (1997) *Human Genetics – Uncertainties and the Financial Implications Ahead*, Transactions of the Royal Society (B), Vol. 352. (Summary booklet also available from Royal Society.)
22. MacDonald AS (1997) How will improved forecasts of individual lifetimes affect underwriting? *Trans. Roy. Soc. B.* pp. 352.
23. Pokorski RJ (1997) Insurance underwriting in the genetic era. *Am. J. Hum. Genet.* **60**: 205–216.
24. Brackenridge RDC, Elder WJ (1992) *Medical Selection of Life Risks*, 3rd Edn. Stockton Press, New York.

Chapter
5

'OVER THE COUNTER' GENETIC TESTING

Lessons from cystic fibrosis carrier screening

P.S. Harper

INTRODUCTION

The discussion of different aspects of genetic testing in the previous chapters has made the basic assumption that there would be some involvement of professionals in the process, and that genetic testing, at least for tests involving serious disorders, would be seen as a medical service rather than simply as a commodity. Although the delivery of genetic testing services may vary between countries according to whether it is provided by state-funded or commercially based laboratories, these have in common, systems of professional control over the way in which testing is requested, while the actual results generated will also be given to the individuals concerned by a medical or genetic professional.

This chapter explores the issues raised by what has come to be known as 'over the counter' testing – the direct provision of services, usually by a commercial company, to an individual, without the request or its result involving any intermediaries at all. Testing of this type is not entirely new, being available already for detection of pregnancy and, in some countries for human immunodeficiency (HIV) status. The possibility of its use for genetic disorders, however, is a new area and one that could have very extensive future implications.

My initial involvement with this topic arose from its unexpected introduction for the testing of carrier status for cystic fibrosis (CF). The following editorial, written for the *Journal of Medical Genetics* shortly after the event, covers the main issues that are still relevant, and I have added a postscript that brings the topic up to date with the events of the subsequent 2 years.

DIRECT MARKETING OF CYSTIC FIBROSIS CARRIER SCREENING: COMMERCIAL PUSH OR POPULATION NEED?[a]

In mid-December 1994, the medical genetics community, along with others, was surprised and concerned by a press and television release announcing the direct marketing of CF carrier screening to the public [1]. The test offered, based on the detection of four common mutations and using DNA from a mouthwash sample sent by mail, also included an information sheet and the possibility of further advice to those concerned at no extra cost. So far, it is too early to know what response and uptake there may have been for the test, but there was certainly considerable coverage of the event in the press and other media.

So why the surprise and why the concern among professionals? The topic of CF carrier screening has, after all, been under intensive study and debate for several years, with a number of pilot studies of different approaches published and in progress [2–8]. As the commonest recessively inherited disorder in northern Europe, with a carrier frequency of 1 in 20–25, it was always likely to attract more commercial interest than rarer conditions and, indeed, industry had already been prominent in developing mutation detection methods appropriate for mass screening. Was not the present development obvious and inevitable?

Why were UK professionals surprised? First, it is probable that this would have been less had the announcement come from the USA, where entrepreneurial involvement in biotechnology and a 'direct sell' approach in medicine generally has always been more common than in the UK and most other European countries, and where a systematic provision of primary care for the entire population is lacking. But a further reason lies in the prominence of UK genetic centres in evaluating CF carrier screening, with pilot projects having already critically assessed its possible benefits and problems in the context of primary care [2,3], pregnancy [6,8] and extended family or 'cascade' screening [9]. It is probably fair to state that, in the context of such extensive and thorough evaluation, those involved with CF carrier screening did not expect a development that would bypass not just themselves but all clinicians, and that this would be introduced with no previous consultation. The final reason for surprise was that 4 months previously, another press release from the same company had indeed announced the development [10], but had specifically stated it would be launched "for use at Medical Genetics Centres and in other clinical laboratories". The possibility of direct marketing to the public was at that time firmly denied.

These facts may help to explain why those involved in the field were both surprised and upset at lack of consultation, but does this really matter? If CF carrier screening is really so beneficial and important, why should professionals object to its more general and direct availability, given that in Britain only a minority of regions have implemented it as a service available as part of overall health care? Here the issues become more complex and illustrate the possible dangers of genetic screening programmes that are driven by commercial considerations rather than grounds of 'health benefit' in the broadest sense.

First, the pilot programmes, especially those based outside pregnancy, have been far from unanimous in their conclusions about the benefit of CF population carrier screening. All agree that it can be done without causing major stress or other harm, provided that the personnel concerned are trained and skilled in giving information about the test and about results, but that this is very different from the situation where the test is delivered in the context of inexpert, or indirectly given information. The Nuffield Foundation report on 'Ethics of Genetic Screening' [11] used CF carrier screening extensively as the basis for its more general conclusion that such screening should only be introduced along with a comprehensive framework of information that allowed fully informed consent and decision-making. It seems unlikely that the ancillary information and counselling service to be provided by the commercial development will in any way fulfil this role, apart from providing help for those concerned by receiving an abnormal result.

Secondly, most pilot studies have shown a marked difference between uptake when testing is offered personally or by mail, raising the question as to whether most of the 'demand' is spontaneous or induced. Since commercial success will equate with numbers of people tested, will there be pressure for testing to be maximized, in terms of the way information is written? Will professionals be induced to promote screening, regardless of individual benefits? This is a further reason for concern in a commercially driven situation.

Thirdly, what is the potential for harm in such direct marketing? The laboratory quality is, it should be said, not in doubt here (though only four mutations are tested for) but how many people will misinterpret a normal result as total reassurance, or an abnormal result as likely to affect their own health? Perhaps the most worrying aspect is that we may never know the answer, though referrals to expert centres to 'pick up the pieces' could increase and should be carefully monitored. Some specific concerns arise: will children be tested? If the policy is no, how will the interests of minors or other third parties be safeguarded? Will abnormal results prejudice insurance? Misinterpretation of this type cannot be ruled out. How will the testing company know that a sample is inappropriate for the screen because of family history or ethnic origins? Disclaimers in the information sheet may avoid legal redress, but will not necessarily help the person involved. All these are issues that can at least be addressed and monitored in a systematic screening service that involves information and counselling as an integral part of the test, but not in a test that is based on direct marketing.

Finally, the main reason why professionals involved in genetics are concerned is the precedent that this development sets for the more general future of genetic testing. This is an area where the dangers, as well as the benefits, have been and are being well researched, especially for such late-onset disorders as Huntington's disease, but also for broader groups such as familial cancers. All this work has emphasized the need for information and genetic counselling to be regarded as an integral part of the test and for it to be resourced as such, if harm is not to outweigh benefit. Will these conclusions be discarded in the face of commercial pressures and the attraction of possible economies from a purely laboratory approach to testing? Will genetic testing be offered directly to the public for such disorders as breast cancer or Alzheimer's disease, where public concern and possible demand is likely to be as high as or higher than for CF? As screening becomes increasingly feasible for common genetic or partly genetic conditions, those working in genetics on both sides of the Atlantic see these threats as real and have repeatedly pointed them out. Will governments need to legislate to control the use and abuse of genetic screening? Those who have launched commercial CF carrier screening to the public may have done us all an unwitting service, if this development provides the focus for effective regulation, so that genetic testing can be harnessed and used responsibly for the benefit of us all.

POSTSCRIPT

The general topic of 'over the counter' genetic testing is one which has received considerable debate since this paper was written in 1995, and the situation is a fluid one in 1997. At present though, it is very much a question of what might happen rather than what has happened already.

Regarding CF, it is extremely relevant that very little has happened. Over the 2 years since University Diagnostics launched its direct marketing of CF screening, there have been no reports of serious problems, nor is there information on the demand during this time, but this is thought to have been small. Advertising has mostly been in women's magazines and has so far avoided targeting pregnant women.

It is also relevant that the demand for CF carrier screening offered through medical channels has so far been limited. In the UK it has only been taken up by very few regions on a service basis, while several of the units who undertook pilot research projects have discontinued them. This has not been because of particular problems arising, but because it was felt that CF carrier screening did not rank as a high priority by comparison with other areas of genetic service, especially in a situation of limited resources. The same pattern has been experienced in the USA where it was widely anticipated as being in high demand but where this has simply not been the case.

Thus, for CF carrier screening, the answer to the question posed originally by

Bekker *et al.* [3] seems to be that it has been more 'supply push' than 'demand pull'. In particular the experience has been that although a high proportion of people accept it if offered immediately face-to-face, only a small proportion actively seek it or take up an offer that involves an interval before testing actually occurs. This raises the question as to what advertising is appropriate to stimulate demand in the face of a largely apathetic public? If commercial priorities are dominant, then one might expect pressures to increase uptake, as would be the case with any 'product' being marketed. This immediately raises ethical and practical issues of free and informed choice and consent, the protection of which is currently not guaranteed or regulated in any way.

Apart from CF carrier screening, no other 'over the counter' genetic testing programmes have currently been introduced in either Britain or America. There is, however, an allied area that deserves looking at where such testing is directly available, 'over the counter', which is the use of DNA 'fingerprinting' for paternity testing. This has now been in operation for over 5 years, with apparent satisfaction, or at least no obvious major problems, and most medical genetics services have been relieved not to have had to be responsible for this non-medical, though sensitive field. This programme also reminds us of the fact, discussed in Chapter 1, that not all genetic testing is necessarily medical in nature.

The other genetic disorder where 'over the counter' testing seems to be an imminent possibility, at least in America, is for familial breast cancer, using the detection of mutations in the susceptibility genes *BRCA1* and *BRCA2*. The American company Myriad Genetics, responsible for the eventual isolation of the *BRCA1* gene, has strongly promoted its use for women at risk, though whether the service will be entirely 'over the counter' or will be on demand through a medical person is not yet clear. So far there has been no comparable development in other countries, though there could be no effective barrier to people using a service marketed directly from America.

If genetic testing for disorders such as familial breast cancer were indeed to be promoted 'over the counter', then many serious issues would arise that are not present for carrier screening in cystic fibrosis. The test result would have direct and serious implications for the individual's future health, for that of relatives, and for the utilization of other health services such as radiology, surgery, and a wide range of clinical and laboratory facilities that might be requested for diagnosis, prevention and treatment, varying according to the nature of the condition. Who is expected to pay for all of these, quite apart from the genetic services likely to be required for family members? It seems most unlikely that the commercial organization promoting the primary genetic test would do so, and the probable scenario would be for the National Health Service or equivalent bodies to be left to 'pick up the pieces' – and to pay for them. It could be argued that the best way to ensure that 'over the counter' genetic testing is not misused is to ensure that the primary testing organization is required to cover the consequential financial costs, which will often be considerably more expensive than the original test.

At present, the 'over the counter' testing being offered for CF carrier state comes with a full explanatory leaflet and with the option of a genetic consultation at no extra cost for those found to be carriers. It is not known to what extent this option has been used or whether the written information given has proved satisfactory; it is quite possible that this may be so for CF carrier testing, but the situation is likely to be different for tests that have major health implications.

At present, 'over the counter' genetic testing is completely unregulated in the UK, so it is perhaps not surprising that it is the first major issue to have been addressed by the newly established Advisory Committee for Genetic Testing. This has issued a consultation document on the topic [13], attracting a considerable response from professional bodies and also from the public through its display on the Internet. A definitive document is expected to be issued in late 1997. Although this body has no direct legal powers, its high profile and official status, together with the possibility of legislation should voluntary regulation fail, make it very likely that it would not be ignored by commercial bodies in the field of genetic testing.

If I am asked to give a personal view as to what place 'over the counter' genetic testing should have in the future pattern of genetic services, I would say that it should be extremely limited. I see no problems with its use in non-medical situations, such as paternity testing; it may be acceptable for the small number of common childhood recessive disorders where the carrier frequency is high and where the carrier state carries no health consequences, but even here I would expect the regular health services to assume responsibility for this if there is clear benefit involved. For the many serious genetic disorders which currently fall under the remit of an overall medical genetics service, I can see neither need for, nor benefit from 'over the counter' testing. Hopefully clear and firm official guidelines, including the need for the cost of testing to include all consequential activities, should help to ensure that it remains a negligible element in our provision of services for genetic diseases.

REFERENCES

1. University Diagnostics. Press release, 12 December 1994.
2. Watson EK, Mayall E, Chapple J *et al.* (1991) Screening for carriers of cystic fibrosis though primary health care services. *Br. Med. J.* **303**: 504–507.
3. Bekker H, Modell M, Denniss G *et al.* (1993) Uptake of cystic fibrosis testing in primary care: supply push or demand pull? *Br. Med. J.* **306**: 1584–1586.
4. Harris H, Scotcher D, Hartley N *et al.* (1993) Cystic fibrosis carrier testing in early pregnancy by general practitioners. *Br. Med. J.* **306**: 1580–1583.
5. Tamber ES, Bernhardt BA, Chase GA *et al.* (1994) Offering cystic fibrosis carrier screening to an HMA population: factors associated with utilisation. *Am. J. Hum. Genet.* **55**: 626–637.

6. Wald NJ (1991) Couple screening for cystic fibrosis. *Lancet*, **338**: 1318–1319.
7. Wald NJ, George LM, Wald WW (1993) Couple screening for cystic fibrosis. *Lancet*, **342**: 1307–1308.
8. Livingstone J, Axton RA, Gilfillan A *et al.* (1994) Antenatal screening for cystic fibrosis: a trial of the couple model. *Br. Med. J.* **308**: 1459–1462.
9. Super M, Schwartz MJ, Malone G, Roberts T, Haworth A, Dermody G (1994) Active cascade testing for carriers of cystic fibrosis gene. *Br. Med. J.* **308**: 1462–1468.
10. Zeneca Group. Press release. 7 August 1994.
11. Nuffield Council on Bioethics (1993) *Genetic Screening. Ethical Issues.* Nuffield Council on Bioethics, London.
12. Andrews LB, Fullarton JE, Holtzman NA, Motulsky AG (1994) *Assessing Genetic Risks. Implications for Health and Social Policy.* National Academy Press, Washington, DC.
13. Advisory Committee for Genetic Testing (1996) *Consultation Paper: Draft Code of Practice for Human Genetic Testing Offered Commercially Direct to the Public.* (Final document to be published September 1997.) HMSO, London.

Section II

GENETIC SCREENING

The second section of this book examines some of the issues raised by genetic screening programmes. The difference between family-based genetic counselling on the one hand, and population-based genetic screening on the other hand is emphasized; the former entails responding to concerns that are already there in a family, while the latter involves raising concerns that may not otherwise exist. One theme recurring throughout this section is the potential for harm caused by a programme of genetic services whose focus is on the population rather than on individuals. This theme is also relevant to later sections when the outcomes of genetic services are discussed in Chapter 12 and the potential abuse of human genetics is discussed in Section IV.

The section begins with an examination of screening for carriers of genetic disease (Chapter 6). For recessive diseases, at least, there are many more carriers than there ever would be parents of affected children. In evaluating such screening programmes, therefore, it is necessary to weigh the potential benefits to a few of making informed reproductive decisions against the lesser potential disadvantages to many more individuals of being identified as carriers. Chapter 7 considers prenatal genetic screening, where some of the potential disadvantages of screening are felt by the whole society because of its effects on every woman's experience of pregnancy as such – it is not necessary for someone to participate in prenatal screening for it to affect them. Prenatal screening programmes also have effects on people with the conditions that they are designed to prevent, for which screening is being offered. How can these, usually invisible, consequences of prenatal screening be taken into account when health services decide for what conditions to make screening available?

Newborn screening programmes are considered in Chapter 8. These are usually regarded as 'the acceptable face of genetic screening' because most newborn screening programmes are intended to provide early treatment for affected infants. Newborn screening for a fatal and incurable disorder, such as Duchenne muscular dystrophy (DMD), does not fulfil the standard, WHO criteria for a screening programme – but there may be good grounds for providing some such screening of newborn infants, subject to certain safeguards, because there may be benefits to the family unit of an early diagnosis. These benefits to the family may also, albeit indirectly, result in benefits to the affected child. Our own programme of newborn screening for DMD is still undergoing a careful evaluation of its social impact.

The final chapter in this section, Chapter 9, examines the issues that may be raised by future genetic screening for susceptibility to the common, complex disorders such as heart disease, diabetes, cancer and dementia. While there may well be therapeutic benefits in the long term from genetic studies of these conditions, there are potential dangers if biotechnology companies attempt to exploit such progress commercially through promoting widespread susceptibility testing of individuals.

The tone of this section is one of caution. This could be seen as the automatic, obstructive response of a traditional physician to a population health measure that will challenge his mode of operation. The alternative view, our own, is that caution is appropriate given the damage that can be done to families who participate even in small-scale genetic testing and screening programmes. We fear the consequences of an uncritical adoption of genetic screening when this is motivated by a cost-saving mentality that has lost sight of the human reality of screening for those caught up in it.

Chapter

6

POPULATION SCREENING FOR GENETIC CARRIER STATUS

A.J. Clarke

INTRODUCTION

Many individuals – if not all – carry altered copies of one or more genes that will never cause them any ill-effects but which they may pass on to their children. If we are to consider testing individuals to identify those who carry such altered ('faulty') copies of particular genes, then we must first consider the two principal biological categories of carrier status, and at least two of the different social contexts in which such testing may be made available.

BIOLOGICAL CATEGORIES OF 'CARRIER'

An unaffected carrier of a genetic disorder may be at risk of transmitting the condition to their children. If the condition is autosomal recessive, then the child will only be affected if both parents are carriers, and if both pass on their altered copy of the gene to the same child. If the condition is sex-linked recessive, then a carrier female will have an affected son if she transmits the disease gene to a child and if the father transmits a Y chromosome (so that the child is male). If the condition is a balanced chromosomal rearrangement (usually a reciprocal translocation), then the carrier parent, of either sex, may hand on an unbalanced set of chromosomes to the child – an extra copy of some chromosome sequences and a missing copy of other sequences. This will usually cause problems with the child's physical and mental development.

There is another sense in which the term 'carrier' can be applied, and that relates to late-onset or variable autosomal dominant disorders which have not (yet) manifested in a particular individual. A person may, for example, be a carrier of the Huntington's disease mutation without (so far) being affected by it; similarly, a person may carry a mutation in the *BRCA1* (breast cancer susceptibility) gene without ever developing a cancer of the breast or any other tissue. The issues relating to testing to identify those at risk of such late-onset or incompletely penetrant disorders – genes which have not yet affected the individual or may never do so – will not be considered further here because they are examined elsewhere in this book (see Chapter 3). The focus in this chapter is on testing for carrier status that is relevant to reproduction only, and not directly to the health of the individual carrier.

In summary, a carrier of a genetic disease may be at risk of having an affected child *either* irrespective of the genetic constitution of their partner (in the case of sex-linked conditions, late-onset or variable dominant disorders and chromosome rearrangements), *or* only if their partner is also a carrier of the same condition (in the case of autosomal recessive diseases). These two types of carrier status need to be distinguished. A note on inheritance is included at the end of this chapter.

SOCIAL CONTEXTS OF CARRIER TESTING

The social contexts in which carrier testing may be available also need to be considered. When a child is affected by a serious genetic disease and family members seek genetic counselling, they will have some knowledge of the condition in practice and they are generally actively seeking information for themselves. Testing to identify carriers in this context is made available through clinical genetics services in response to family requests, and the individuals in each family are usually able to consider and discuss the possible consequences for themselves and others of their being tested or not being tested. This type of genetic testing is offered in the context of thoughtful and supportive counselling, and is tailored to the individuals and families involved.

In contrast, the other context in which carrier testing may be made available is that of general population screening. This is usually a pro-active programme, initiated by professionals and promoted by health services. Many individuals will be approached and offered testing for a condition of which they probably know little and have no direct experience. It is also the professionals who will have decided for which diseases screening should be offered. It is professionals who will have decided which individuals to approach and in what setting, and they will decide what information to provide and what opportunities to allow for reflection and for consideration of whether or not to have the testing carried out. In the context of population screening, therefore, it may be more difficult to ensure that the needs of those offered testing for information and counselling before making a decision have been recognized, that the offer of testing is made in an appropriate manner, and that screening does not cause more problems than it is likely to solve.

It should perhaps be pointed out, that the contrast I have drawn here between biological and social categories of testing is not as clear-cut as I may have suggested. The two 'biological' categories have very different social consequences for identified carriers, and of course the distinction between the two 'social' categories in fact has a biological basis defined on the basis of the probability of a person tested being a carrier.

POSSIBLE PROBLEMS FROM CARRIER TESTING

Testing for carriers of an autosomal recessive disease will identify many more carriers than carrier couples. One person in 25 in Britain carries a disease-causing copy of the cystic fibrosis (CF) gene. One in four people in some parts of West Africa carries the gene for sickle-cell disease and is thereby relatively resistant to certain types of malaria. This indicates the difficulty in labelling some alleles (copies, versions) of a gene as 'faulty' because the sickle cell mutation is clearly advantageous in this context, although it may cause serious problems to those with two copies of the sickle cell gene. If the process of testing and being identified as a carrier causes anxiety, confusion or concerns in even a modest proportion of individuals, then this must be weighed against any advantages experienced by the relatively few couples who will be at risk of an affected child and therefore in a position to make use of the information in their reproductive planning.

What has been learned so far about problems resulting from carrier testing in practice – in family-based testing and in screening programmes? It is clear that being identified as a carrier can cause significant problems. Stigmatization at the personal, social level has been acknowledged as a problem over many years [1,2]. Furthermore, because gene frequencies differ in various populations, there is the opportunity for genetic testing programmes to exacerbate racial discrimination, as occurred with sickle cell screening in the USA during the 1970s [3,4]. It is still possible for genetic carrier testing to be abused to bolster racism, and to lend pseudo-scientific credibility to claims of ethnic identity and separateness [5,6].

At the personal level, some of those shown to be carriers of a recessive disease can harbour regrets at having been tested or they may have lingering concerns about their carrier status [7,8]. They may have less positive feelings about themselves [2] or feel less optimistic about their own future health than non-carriers [9]. Their memory of their test result and its significance may become confused over time [10,11].

'INFORMED REPRODUCTIVE DECISIONS' AS A BENEFIT?

The principal intended consequence of carrier testing is that it enables carrier couples to make informed reproductive decisions about their future reproduction: whether to try for a pregnancy, whether to carry out testing on a future partner or pregnancy, and whether to continue a pregnancy or terminate it. Some individuals and couples will regard this as a benefit, but others will not. Instead, some will perceive the process of carrier testing, the information that one is a carrier, the disclosure of this to a partner or a potential partner and then the making of 'informed reproductive decisions' by the carrier couple as in fact a series of burdens that they would prefer to have avoided. In the absence of any family history of the disease in question, they might prefer to have simply taken their chances (1 in 2500 for CF in Britain).

Can the making of 'informed reproductive decisions' be regarded as a valid goal of genetic testing or screening? The answer has to depend upon what the testing is for, and what the process of testing entails. It is not possible to regard such decision-making as an unqualified good, or a valid goal in its own right. What the testing is for must be significant, otherwise prenatal selection for sex or on cosmetic criteria would not be so problematic. Of course, any genetic counselling service or screening programme must fully respect the autonomy of its clients while pursuing its goals, but the autonomous, informed decisions of its clients do not themselves amount to such a goal [12,13]. This is an important aspect of carrier screening programmes that is not addressed adequately by Modell and Kuliev [14].

THE POTENTIAL BENEFITS OF INFORMED REPRODUCTIVE DECISIONS

So when would informed reproductive decisions be regarded as a benefit? In principle, it would clearly be all to the good if the inevitable suffering of infants affected by serious genetic conditions, and the distress caused to their families, could be avoided. In practice, this would entail either avoiding the conception of infants likely to be affected or opting for prenatal diagnosis and the selective termination of affected pregnancies. The possible benefits of informed reproductive decisions, then, must be weighed against the possible problems caused by the process and consequences of carrier testing or screening. These include the concerns raised in the many carriers whose partners are not also carriers; the sadness of the decision to remain childless; the practical and emotional difficulties associated with prenatal diagnosis; and the distress caused by the termination of wanted pregnancies. Offence and distress caused to affected individuals in society and to their families by the promotion of screening programmes must not be forgotten.

EXPERIENCE WITH FAMILY-BASED CARRIER TESTING

Attitudes to prenatal diagnosis for CF among the parents, aunts and uncles of affected children is very varied, but only a minority of those questioned state that they would terminate an affected pregnancy [15,16]. It must be remembered that the aunts and uncles of affected children are very likely to know about the condition affecting their niece or nephew, but they may not be aware of the mode of inheritance or its implications for their own carrier status [17]. It does appear, however, that many parents of affected children would avoid further children rather than opt for prenatal diagnosis [18], and the uptake of prenatal diagnosis in families with an affected child has been relatively low – with testing carried out on only four out of 24 pregnancies in one series [19]. If the behaviour of those with personal knowledge and experience of a disease can act in some sense as a guide to the likely behaviour of others once they have been adequately informed about the particular disease, then such findings could make us rather cautious about the wisdom of actively promoting population screening programmes.

When carrier testing is made available to the adult sibs of individuals with CF, many of these potential carrier siblings choose to be tested but many others do not. These CF siblings frequently experience considerable emotional distress when being tested and when given their test results (whether carriers or not), and many prefer to remain unaware of their genetic status. Indeed, ignorance of their carrier status can serve the function of avoiding difficult topics within the family and can protect the sibling from a perceived loss of social desirability [20,21]. These processes are comparable to the joint approach of health professionals and CF patients to their discussions about the course of the illness, when they soften their descriptions of the disease and thereby enable both groups to cope with the uncertainties of a chronic, terminal disease [22].

Experience of carrier testing in the context of X-linked and chromosomal disorders has also demonstrated the complexity of individual responses to information about their genetic 'risk' status – their risk of being carriers or of having children affected by a serious condition. In families with Duchenne muscular dystrophy (DMD), young women who are or could be carriers of this sex-linked disease will usually not dwell upon this topic, but their risk status will intrude into their lives at particular times – such as when they begin a potentially significant relationship with a boyfriend. Furthermore, their understanding of their risk status (often given by the professionals as a percentage chance that they are carriers of DMD) will tend to be polarized towards 'definitely a carrier' or 'definitely not a carrier', in their memory and reasoning [23,24]. Such work has shown that the decision-making process in relation to reproductive behaviour is not so much a rational calculus in the face of known (objective and quantified) risks, but depends also upon a variety of other personal and social factors such as the strength of the desire for children [25,26], which will in turn depend upon the person's relationship with their partner.

In families with balanced chromosomal rearrangements, we know that the diffusion of information about the diagnosis within families can be variable, with some families or family members resisting the process [27], and we know that parents often find it difficult to discuss the issue of carrier status with their children [28]. This is also a problem in the context of CF [21].

It can be appreciated that genetic carrier testing can be associated with emotional distress and psychosocial problems, even within families in which there is a pre-existing awareness of genetic risk. Because the problems and the awareness of risk are present independently of the genetic counselling, however, these difficulties confirm the appropriateness of providing facilities for genetic counselling and testing in such families rather than argue against it.

EXPERIENCE WITH POPULATION CARRIER SCREENING

Studies of population carrier screening programmes for CF show a different pattern. Several studies have shown that testing can be offered in a primary care setting without causing reports of serious distress or clinically important levels of anxiety or depression. Understanding of the test results has generally been adequate, although long-term recall and understanding of the information provided may not be so good [11].

When such carrier screening is made available to the general adult (non-pregnant) population, the rate of uptake of the test is very variable. More than anything else, the rate of uptake depends upon the manner in which the offer of testing is made. When the offer has been made by letter, the uptake in several British studies has been of the order of 10–15% [29–31]. When the offer has been made opportunistically and in person by the researcher when the subjects were attending their primary health care centre for other reasons, and was for on-the-spot testing, the uptake was around 65–70% (uptake was somewhat higher in a Family Planning Clinic setting [29]). In contrast, if the offer was for testing on another day the uptake in one study was 25%, and if the offer was made on a leaflet handed to the patient by the receptionist on arrival the uptake was 17% [30]. These findings have been interpreted as revealing a general lack of motivation to be tested but a willingness to comply with the advice of respected professionals – apathy and compliance in the face of professional enthusiasm. A similar pattern of results, although a lower rate of uptake, was found in a study within a Health Maintenance Organization practice in the USA [32].

When CF carrier testing is offered in pregnancy, the rate of uptake is generally higher (74–87%) [33,34] and is nearly 100% when the test is offered early in pregnancy by the woman's family doctor [35,36]. High rates of uptake in antenatal clinics have also been found in Germany [37] and Denmark [8], and comparable findings are emerging from studies in the USA [38]. It is not clear whether the uptake is higher in a pregnancy because parents perceive the test as more relevant to them at that time or because women find it more difficult to decline the offer of such tests in pregnancy. They certainly do find it difficult to decline other genetic screening tests in pregnancy [39], and prenatal genetic screening tests are frequently offered in a frankly directive, prescriptive manner [40,41].

Given that the mode of offering a test influences uptake, and that women find it hard to decline the offer of prenatal tests, the active promotion and marketing of CF carrier screening by the private, commercial sector, could well be very 'effective' in achieving high levels of uptake. This could lead to real problems for substantial numbers of people, especially those who complied with the suggestion that they be screened without seriously considering in advance the potential disadvantages of being tested. Another factor that may lead to the promotion of carrier screening programmes is professional enthusiasm for the new technologies [42], which may or may not be associated with commercial factors. It should not be forgotten that there is a growing potential for conflicts of interest when physicians become involved in commercial applications of the new genetic technologies. For example, there would be a problem if a clinical geneticist was promoting carrier testing for CF to his patients, and if the DNA test kit was supplied by a corporation in which he (the geneticist) had a consultancy or held shares.

An example of over-enthusiasm occurs in a paper by Watson *et al.* reporting on CF carrier screening in primary care. They concluded that "carriers and non-carriers uniformly approve of screening and are glad to have been tested" [43]. This sentence in the summary contrasts with the data presented in detail in the paper, where only 60% of carriers returned the study's follow-up questionnaire at 6 months and where 70% of the respondents (only 42% overall) stated at 6 months post-test that they were either indifferent or not worried; the rest of the respondents were slightly anxious, worried or depressed. Of the identified carriers who returned the questionnaires, 6% did not state that they were glad to have been tested and 12% were unsure. My interpretation of the authors' summary of their results is not that they were deliberately trying to minimize the possible problems from carrier screening but that their enthusiasm for screening led them to make rather uncautious generalizations that went beyond the evidence. Care must be taken to avoid the uncritical acceptance of sweeping generalizations, especially when these generalizations contain bland reassurance about proposed new developments.

THE SETTING OF CARRIER SCREENING

There has been some discussion about the best way to offer carrier testing for cystic fibrosis, whether to the general population or in the antenatal context. The usual way has been to test individuals, and to suggest that it would be worthwhile testing the partner of an identified carrier. The alternative way, proposed particularly by Wald [44], would be to take samples from both members of a couple and test them as a unit – giving results to the couple as either both shown to be carriers, or not both shown to be carriers (i.e. either one, or neither, shown to be carriers). Such an approach may have some advantages, especially antenatally, in alarming considerably fewer couples than with individual carrier testing, but of course there is the other possibility of not offering carrier screening in the antenatal clinic at all. There are also disadvantages to couple testing, including the deliberate concealment of information generated in the laboratory, the false reassurance to individuals who think they are not carriers and who then have an affected child with another partner, and the inability of many identified carriers' extended families to have the option of genetic counselling and cascade testing. When a group in Aberdeen compared couple testing with the stepwise testing approach of Mennie *et al.* [34], the carriers identified using the stepwise method were anxious until their partners had tested negative, while those given negative results with the couple testing approach had higher persisting levels of residual anxiety [45]. On balance, the Aberdeen group judged that stepwise testing was preferable.

One other context in which carrier testing for autosomal recessive diseases has been offered, is in a high school setting and in association with an educational programme. Experience in Montreal indicates that a high proportion of the High School population can be reached in this way – at least in relation to Tay–Sachs disease, β-Thalassaemia and CF, and that more than half of the students choose to be tested [46,47]. In this setting, it seems that students recall their test results and use them in making reproductive decisions when older. Given the great care that has been put into ensuring that the testing is truly voluntary and confidential (which will be difficult to ensure in many school settings) such an approach to screening linked with education may be helpful. There are difficulties, however, with the interpretation of the school students' generally reassuring response to direct questions about the possible impact of a positive carrier test on self-esteem or on the desirability of a partner. These findings do not amount to proof that no such effects will occur; the same group have previously reported results that were not quite so reassuring [7]. Furthermore, it is not entirely clear to what extent the effects of this programme are the result of the testing itself or of the accompanying educational input.

POTENTIAL PROBLEMS OF POPULATION CARRIER SCREENING

The potential problems arising from population carrier screening programmes may be considered under three categories – the consequences for individuals, the personal–social consequences that relate to relationships between individuals, and the collective–social or institutional consequences. First, the problems arising within the individual, such as the distress it may cause in a person identified as a carrier, and the burden of reproductive responsibility. We have already discussed some of these consequences, and have indicated some aspects of the burden of reproductive decision-making that may arise. Another burden that many practising clinical geneticists will be keen to reduce is the sense of guilt experienced by many parents and grandparents of children who develop a serious genetic disease. How may this be done? Often enough, by stating " . . . There was nothing you could have done to avoid this happening . . . ". But this may not be entirely satisfactory. While this phrase may be true when used today, or next year, will it be true in 10 or 20 years' time? Perhaps, by using this phrase now, we clinicians are inadvertently strengthening the guilt that parents of an affected child may experience in the future, thereby reinforcing the pressure on individuals to accept carrier screening programmes.

If society comes to view genetic disease as potentially avoidable, will the parents of children affected by genetic disorders be socially isolated or stigmatized? Will carriers be stigmatized and regarded by non-carriers as unhealthy or tainted, and therefore as unsuitable partners in relationships? Will this lead to diminished respect for individuals with genetic conditions and indeed for all those with mental or other disabilities? These already are real problems for individuals with Down syndrome, fragile X syndrome and conditions associated with readily visible stigmata; they could be exacerbated by the development of more systematic carrier screening programmes.

At the institutional level, may there be large-scale effects on society? Society may decide that it is unwilling to provide financial or social support for the families of individuals with 'preventable' genetic conditions. Could genetic conditions come to vary in frequency with social class, being highest at the two ends of the spectrum of income distribution or wealth? Such disorders may become relatively more common among the poor, who usually have poor access to secondary and tertiary level health care, and perhaps also among the wealthy, who may feel that they can afford to cope with a child affected by a genetic condition, so that they do not need to accept antenatal screening tests offered in antenatal clinics.

If society decides to maximize its gains from investment in genetic research, then it may promote carrier screening and prenatal screening programmes very actively [48]. In a free-market dystopia, especially where health care is provided predominantly through private insurance schemes, there could be financial penalties for those who decline the offer of a genetic screening test; these could be the loss of specific social service benefits, a possible lack of insurance cover, and heavy medical bills if health-care provision were denied to those children with costly disorders who could have been 'prevented'.

This then raises the fascinating, although perhaps rather fanciful, question of what justifications would be accepted for opting out of screening programmes; perhaps there would be a parallel with pacifism as a justification for opting out of military conscription. 'Religious reasons' would be put forward by at least some anti-testers to justify their decision. Would one have to be a member of a specific church or other faith? For how long? How would others be viewed who found that they could not accept prenatal genetic screening programmes or a termination of pregnancy because of their abhorrence at the prospect of abortion?

COLLECTIVE DECISIONS ABOUT CARRIER SCREENING

How should society come to a decision about what types of carrier testing and screening should be available? There are more practical and ethical difficulties raised by population carrier screening programmes than by carrier testing offered within a family context, so the main questions will naturally focus on issues of screening. Carrier testing within recognized families – active cascade testing – is relatively unproblematic [49], but this is clearly not true for population screening. Should population carrier screening, then, be actively promoted by the health services, permitted by private, commercial enterprises, or prevented by law?

Having recognized 'informed reproductive decisions' as not in itself an adequate reason to set up a screening programme, we must consider the two other principal arguments in favour of carrier screening: the avoidance of the suffering of affected individuals and their families, and the saving of resources that results if the process of carrier testing, prenatal diagnosis and the termination of affected pregnancies is less costly than caring for those affected by the disease under consideration. The interplay between these two types of justification will be illustrated briefly in relation to four different genetic conditions.

Tay–Sachs disease is, at present, essentially untreatable. It is a disease of infancy and early childhood that leads to progressive neurological deterioration and then the death of a young child. It is most common in Ashkenazi Jewish communities, where 1 person in 30 may be a carrier. It causes suffering to the affected children and distress to their families. Because of the inability to treat the disease, it is not very costly in financial terms. β-Thalassaemia, in contrast, is a very treatable disease. If untreated it will often cause a slow physical decline throughout childhood and adolescence, entailing prolonged physical and emotional suffering for the affected persons and their families. It is expensive to treat, however, especially for Mediterranean and Third World countries with a high incidence of the disease but with few resources so that adequate treatment for all those affected would consume much of their national health-care budgets.

CF is common in north-western European populations, and is a cause of severe and long-lasting morbidity and of premature mortality. Calculations about the likely future costs of treatment [50], however, are plagued by uncertainties. Rational therapy based upon an understanding of the function of the CF gene may well improve the long-term outcome for affected individuals, but it is not known whether this will greatly increase or decrease the lifetime costs of treatment. Already, life expectancy for those born today is put at 40 years or more, and the coming of new pharmacological remedies or gene therapy may improve quality of life and life expectancy still further. Such improved treatments may make the prospect of prenatal diagnosis less appealing to families, but if these treatments are costly then screening and prenatal diagnosis may seem more appealing to those planning and funding health services.

Finally, we will consider screening for female carriers of fragile X syndrome – indeed, it has been proposed as a way of reducing the incidence of this relatively common familial form of intellectual disability [51]. While it may be thoroughly appropriate for members of fragile X families to seek genetic counselling and testing, there are technical and ethical difficulties that make it more difficult to justify population carrier screening. The principal technical difficulty is the inability to distinguish between a female fetus carrying fragile X who would be affected, and one who would be intellectually 'normal' and therefore an unaffected carrier. One serious ethical difficulty is that fragile X syndrome is not quite a disease in the usual sense of the word – it is a genetic condition, but it does not cause progressive ill health and does not generally entail physical suffering for those who are affected. Much of the distress caused by the condition is experienced by the family of the affected individual, and it could be argued that this distress, and any suffering of the affected individuals that is caused by processes of stigmatization and social intolerance, could be greatly ameliorated if the support offered by society to these families was improved. A carrier screening programme could even exacerbate the situation by making society less tolerant of affected individuals and their parents. A further ethical difficulty arises from the fact that some female carriers of fragile X syndrome can be affected by the condition – they may have a degree of intellectual disability. In the context of a population screening programme, how realistic would it be to expect that such women will be able to make a truly informed decision to accept the offer of testing; the dangers of routinization undermining their autonomy must be very real.

How then should society weigh up the benefits to be gained from a carrier screening programme against the burdens of imposing genetic decision-making on their populations, together with the dangers of reinforcing stigmatization and even racism that may accompany such programmes? There are no easy answers. Given that health service resources are limited, and that commercial promotion of carrier testing could lead many to be tested who would subsequently regret it, the standard of proof that a screening programme will be of overall benefit must be high, and the possibility of adverse effects must be considered very carefully. In my view, priority should be given at present to screening programmes that will identify affected individuals early to permit improved treatment. For a discussion of these issues in relation to newborn screening for Duchenne muscular dystrophy, see Chapter 8.

CONCLUSIONS: POSSIBLE DIRECTIONS IN CLINICAL GENETICS

It is possible to draw an outline of two possible futures, depending on how our society responds to the facts of genetic disease. These word-sketches reflect the ethos of two different approaches to genetic disorders. One response is to nurture and care for affected individuals, to provide family support as appropriate, and to make available a high-quality family-centred genetic counselling service.

In contrast, another pattern of response to genetic disease is the approach of prevention by programmes of carrier screening and prenatal diagnosis. This could allow society to absolve itself of its ongoing responsibilities for children with mental disability or other genetic conditions, with the responsibility being devolved to the parents of affected children. Those individuals who opt out of screening programmes, or who do not choose to terminate an 'affected' pregnancy, may then be blamed as morally or socially irresponsible. The use of the word, 'prevention', in many health care contexts will lead on to the notion of 'responsibility', and thence to the stirring of the flames of blame and guilt [52]. In the context of genetics, the corresponding notion is 'reproductive responsibility'; the failure to comply with the recommended screening tests is already commonly regarded as blameworthy [53]. It would be completely unacceptable for a health service deliberately to provoke such emotions in the parents of children with serious diseases. It is simply not appropriate to approach prevention in the context of genetic disease in the same way as the health education and anti-tobacco campaigners lobby for the prevention of lung cancer and other ills caused by smoking.

It would be too easy, however, to reject screening and prenatal diagnosis out of hand without retaining an awareness of the real suffering and distress caused by genetic disease and experienced by the affected individuals and their families. Those individuals making reproductive decisions in relation to genetic disease do need to have information available about these conditions that is realistic and not rosy.

Elements of both the nurturing and the preventive approaches are present in contemporary society, and it is unlikely that any future will conform strictly to either model. But such models may serve a purpose in clarifying concepts and values, and thereby helping us to shift society more towards the nurturing model, than the model of prevention riding roughshod over individuals.

REFERENCES

1. Stamatoyannopoulos G (1974) Problems of screening and counselling in the hemoglobinopathies. In: *Birth Defects: Proceedings of the Fourth International Conference*. (eds AG Motulsky, FLB Ebling) Excerpta Medica, Amsterdam.
2. Evers-Kiebooms G, Denayer L, Welkenhuysen M, Cassiman J-J, Van den Berghe H (1994) A stigmatizing effect of the carrier status for cystic fibrosis? *Clin. Genet.* **46**: 336–343.
3. Culliton BJ (1972) Sickle cell anemia: the route from obscurity to prominence. *Science*, **178**: 138–142.
4. Hampton ML, Anderson J, Lavizzo BS, Bergman AB (1974) Sickle cell 'nondisease': a potentially serious public health problem. *Am. J. Dis. Child.* **128**: 58–61.
5. Bradby H (1996) Genetics and racism. In: *The Troubled Helix: Social and Psychological Implications of the New Human Genetics* (eds TM Marteau, MPM Richards), pp. 295–316. Cambridge University Press, Cambridge.
6. Macbeth H (1997) What is an ethnic group? – a biological perspective. In: *Culture, Kinship and Genes* (eds A Clarke, EP Parsons). Macmillan, London (in press).

7. Zeesman S, Clow CL, Cartier L, Scriver CR (1984) A private view of heterozygosity: eight-year follow-up study on carriers of the Tay–Sachs gene detected by high school screening in Montreal. *Am. J. Med. Genet.* **18**: 769–778.
8. Clausen H, Brandt NJ, Schwartz M, Skovby F (1996) Psychological impact of carrier screening for cystic fibrosis among pregnant women. *Eur. J. Hum. Genet.* **4**: 120–123.
9. Marteau TM, van Duijn M, Ellis I (1992a) Effects of genetic screening on perceptions of health: a pilot study. *J. Med. Genet.* **29**: 24–26.
10. Loader S, Sutera CJ, Segelman SG, Kozyra A, Rowley PT (1991) Prenatal hemoglobinopathy screening. IV. Follow-up of women at risk for a child with a clinically significant hemoglobinopathy. *Am. J. Hum. Genet.* **49**: 1292–1299.
11. Axworthy D, Brock DJH, Bobrow M, Marteau TM (1996) Psychological impact of population-based carrier testing for cystic fibrosis: 3-year follow-up. *Lancet*, **347**: 1443–1446.
12. Chadwick R (1993) What counts as success in genetic counselling? *J. Med. Ethics*, **19**: 43–46.
13. Clarke A (1993) Response to: What counts as success in genetic counselling? *J. Med. Ethics*, **19**: 47–49.
14. Modell B, Kuliev AM (1993) A scientific basis for cost–benefit analysis of genetics services. *Trends Genet.* **9**: 46–52.
15. Denayer L, Evers-Kiebooms G, De Boeck K, Van den Berghe H (1992a) Reproductive decision-making of aunts and uncles of a child with cystic fibrosis: genetic risk perception and attitudes towards carrier identification and prenatal diagnosis. *Am. J. Med. Genet.* **44**: 104–111.
16. Wertz DC, Janes SR, Rosenfield JM, Erbe RW (1992) Attitudes toward the prenatal diagnosis of cystic fibrosis: factors in decision-making among affected families. *Am. J. Hum. Genet.* **50**: 1077–1085.
17. Denayer L, De Boeck K, Evers-Kiebooms G, Van den Berghe H (1992b) The transfer of information about genetic transmission to brothers and sisters of CF parents with an affected CF-child. *Birth Defects Original Article Series*, **28 (1)**: 149–158.
18. Watson EK, Marchant J, Bush A, Williamson R (1992) Attitudes towards prenatal diagnosis and carrier screening for cystic fibrosis among the parents of patients in a paediatric cystic fibrosis clinic. *J. Med. Genet.* **29**: 490–491.
19. Jedlicka-Kohler I, Gotz M, Eichler I (1994) Utilization of prenatal diagnosis for cystic fibrosis over the past seven years. *Pediatrics*, **94**: 13–16.
20. Fanos JH, Johnson JP (1995) Perception of carrier status by cystic fibrosis siblings. *Am. J. Hum. Genet.* **57**: 431–438.
21. Fanos JH, Johnson JP (1995) Barriers to carrier testing for adult cystic fibrosis sibs: the importance of not knowing. *Am. J. Med. Genet.* **59**: 85–91.
22. Waddell C (1982) The process of neutralisation and the uncertainties of cystic fibrosis. *Sociol. Hlth Illness*, **4**: 210–220.
23. Parsons EP, Atkinson P (1992) Lay constructions of genetic risk. *Sociol. Hlth Illness*, **14**: 437–455.
24. Parsons EP, Clarke A (1993) Genetic risk: women's understanding of carrier risks in Duchenne muscular dystrophy. *J. Med. Genet.* **30**: 562–566.
25. Frets PG, Duivenvoorden JH, Verhage F, Ketzer E, Niermeijer MF (1990) Model identifying the reproductive decision after genetic counselling. *Am. J. Med. Genet.* **35**: 503–509.

26. Parsons EP, Atkinson P (1993) Genetic risk and reproduction. *Sociol. Rev.* **41**: 679–706.
27. Ayme S, Macquart-Moulin G, Julian-Reynier C, Chabal F, Giraud F (1993) Diffusion of information about genetic risk within families. *Neuromusc. Disord.* **3**: 571–574.
28. Jolly A, Parsons EP, Clarke A (1996) Testing children to identify carriers of balanced chromosomal translocations: a retrospective, qualitative psychosocial study. Abstract 9.101 of 28th Meeting of European Society of Human Genetics, London, April 1996, *and* British Human Genetics Conference, York, September 1996. *J. Med. Genet.* **33** (Suppl 1), abstract 6.018.
29. Watson EK, Mayall E, Chapple J, Dalziel M, Harrington K, Williams C, Williamson R (1991) Screening for carriers of cystic fibrosis through primary health care services. *Br. Med. J.* **303**: 504–507.
30. Bekker H, Modell M, Denniss G, Silver A, Mathew C, Bobrow M, Marteau T (1993) Uptake of cystic fibrosis testing in primary care: supply push or demand pull. *Br. Med. J.* **306**: 1584–1586.
31. Payne Y, Williams M, Cheadle J *et al.* (1997) Carrier screening for cystic fibrosis in primary care: evaluation of a project in South Wales. *Clinical Genetics*, **51**: 153–163.
32. Tambor ES, Bernhardt BA, Chase GA *et al.* (1994) Offering cystic fibrosis carrier screening to an HMO population: factors associated with utilization. *Am. J. Hum. Genet.* **55**: 626–637.
33. Wald NJ, George LM, Wald NW (1993) Couple screening for cystic fibrosis. *Lancet*, **342**: 1307–1308.
34. Mennie ME, Gilfillan A, Compton M, Curtis L, Liston WA, Whyte DA, Brock DJH (1992) Prenatal screening for cystic fibrosis. *Lancet*, **340**: 214–216.
35. Harris H, Scotcher D, Hartley N, Wallace A, Craufurd D, Harris R (1993) Cystic fibrosis carrier testing in early pregnancy by general practitioners. *Br. Med. J.* **306**: 1580–1583.
36. Harris H, Scotcher D, Hartley N, Wallace A, Craufurd D, Harris R (1996) Pilot study of the acceptability of cystic fibrosis carrier testing during routine antenatal consultations in general practice. *Br. J. Gen. Pract.* **46**: 225–227.
37. Jung U, Urner U, Grade K, Coutelle C (1994) Acceptability of carrier screening for cystic fibrosis during pregnancy in a German population. *Hum. Genet.* **94**: 19–24.
38. Loader S, Caldwell P, Kozyra A, Levenkron JC, Boehm CD, Kazazian HH, Rowley PT (1996) Cystic fibrosis carrier population screening in the primary care setting. *Am. J. Hum. Genet.* **59**: 234–247.
39. Sjogren B, Uddenberg N (1988) decision-making during the prenatal diagnostic procedure. A questionnaire and interview study of 211 women participating in prenatal diagnosis. *Prenat. Diagn.* **8**: 263–273.
40. Marteau T, Slack J, Kidd J, Shaw R (1992b) Presenting a routine screening test in antenatal care: practice observed. *Publ. Hlth*, **106**: 131–141.
41. Marteau TM, Plenicar M, Kidd J (1993) Obstetricians presenting amniocentesis to pregnant women: practice observed. *J. Reprod. and Infant Psychol.* **11**: 3–10.
42. Koch L, Stemerding D (1994) The sociology of entrenchment: a cystic fibrosis test for everyone? *Soc. Sci. Med.* **39**: 1211–1220.
43. Watson EK, Mayall E, Lamb J, Chapple J, Williamson R (1992) Psychological and social consequences of community carrier screening programme for cystic fibrosis. *Lancet*, **340**: 217–220.

44. Wald NJ (1991) Couple screening for cystic fibrosis. *Lancet*, **338**: 1318–1319.
45. Miedzybrodzka ZH, Hall MH, Mollison J, Templeton A, Russell IT, Dean JCS, Kelly KF, Marteau TM, Haites NE (1995) Antenatal screening for carriers of cystic fibrosis: randomised trial of stepwise vs couple screening. *Br. Med. J.* **310**: 353–357.
46. Mitchell J, Scriver CR, Clow C, Kaplan F (1993) What young people think and do when the option for cystic fibrosis carrier testing is available. *J. Med. Genet.* **30**: 538–542.
47. Mitchell J, Capua A, Clow C, Scriver CR (1996) Twenty-year outcome analysis of genetic screening programs for Tay–Sachs and β-Thalassaemia disease carriers in high schools. *Am. J. Hum. Genet.* **59**: 793–798.
48. Clarke A (1990) Genetics, ethics and audit. *Lancet*, **335**: 1145–1147.
49. Super M, Schwarz MJ, Malonc G, Roberts T, Haworth A, Dermody G (1994) Active cascade testing for carriers of cystic fibrosis gene. *Br. Med. J.* **308**: 1462–1468.
50. Cuckle HS, Richardson GA, Sheldon TA, Quirke P (1995) Cost effectiveness of antenatal screening for cystic fibrosis. *Br. Med. J.* **311**: 1460–1464.
51. Palomaki GE (1994) Population based prenatal screening for the fragile X syndrome. *J. Med. Screening*, **1**: 65–72.
52. Sachs L (1996) Causality, responsibility and blame – core issues in the cultural construction and subtext of prevention. *Sociol. Hlth Illness*, **18**: 632–652.
53. Marteau TM, Drake H (1995) Attributions for disability: the influence of genetic screening. *Soc. Sci. Med.* **40**: 1127–1132.

APPENDIX: A NOTE ON INHERITANCE

Genes are the coded instructions that control the growth and development of the body. There are many thousands of different genes which come in matching pairs – one from each parent. Chromosomes are physical structures that carry the genes, rather like beads on a length of string. Chromosomes are important in regulating gene expression and in ensuring the correct transmission of genes through cell divisions. Most genes are present in two copies in each individual, with one copy coming from each parent. This is true for the genes on the 22 pairs of chromosomes which are the same in men and women, the autosomes. The other pair of chromosomes are the sex chromosomes, X and Y; a girl has two X chromosomes and a boy has one X chromosome and one Y chromosome. The X chromosome carries many genes apart from those involved in sex and reproduction; the Y chromosome makes males male but does little else.

A genetic disease will result if gene function is disrupted by mutation. If just one functioning copy of a gene is sufficient to avoid problems, then the disease caused by a lack of function of both copies is termed recessive; the 'faulty' copy of the gene is masked by the intact copy. If two carriers of the same altered gene have a child, and if both transmit their altered copy of the gene to the same child, then that child will be affected by the disease. If one child in a family is affected, there will be a 1:4 (25%) chance of recurrence in another child.

If a single faulty copy of the gene is sufficient to cause disease, then the disease will be inherited as a dominant trait – an affected person has a 1:2 (50%) chance of transmitting the condition to any child.

If a faulty gene on the X chromosome is present in a male, he will usually be affected by the corresponding sex-linked disease because he only has a single copy of that gene; if the faulty copy is present in a female then she will usually be unaffected, although the pattern of X chromosome inactivation may result in her manifesting some signs of the disorder (only one X chromosome is active in any cell in the woman's body, and the decision as to which X chromosome to inactivate is made in each cell at some point in the early embryo).

Chapter

7

THE GENETIC DISSECTION OF MULTIFACTORIAL DISEASE

The implications of susceptibility screening

A.J. Clarke

INTRODUCTION

The techniques of molecular genetics have been applied to human genetic diseases for nearly two decades, and have led to the identification of numerous genes implicated in the causation of single-gene (Mendelian) disorders. The genetic basis of multifactorial traits and disorders has also been investigated, but has been slower to yield to the process of inquiry. Multifactorial disorders have been thought to arise from the interaction in an individual of multiple genes, each generally of minor effect, with the modifying influence of environmental factors. Most of the common degenerative, neoplastic and psychiatric disorders of Western society are regarded as multifactorial in aetiology, and polygenic in so far as they are inherited.

By studying families with more than one affected individual, to determine what portions of the genome are shared between them, it has been possible to measure the contribution of different genetic loci to the pathogenesis of several multifactorial disorders. In the case of diabetes mellitus type 1 (juvenile-onset, insulin-dependent diabetes), genetic susceptibility is influenced by more than a dozen loci including one important locus within the major histocompatibility complex on chromosome 6; this locus is responsible for some 42% of the familial clustering found in this disease [1,2]. Changes in the mitochondrial genome as well as in nuclear genes may influence susceptibility to this and other such conditions.

This chapter is based on a paper by A.J. Clarke published in the *British Medical Journal*, Vol. 311, pp. 35–38, and appears here by kind permission of the BMJ Publishing Group.

In Hirschprung's disease, another disorder traditionally regarded as 'multifactorial', it has become clear that a significant number of cases – of sporadic and familial disease – are accounted for by mutations in the RET oncogene. Other cases are accounted for by mutations in the endothelin receptor B gene. While there is no single cause of this disease, it does seem that in any affected individual it will often be the result of a change in one of several distinct, virtually Mendelian genes. A genetic basis for susceptibility to certain malformations is also emerging, as with orofacial clefting and neural tube defects. It appears that a small sub-group of these malformations arises as a result of defects at single, Mendelian loci, and that a larger proportion results from interactions between genetic and environmental factors; dietary intake of folic acid, for example, appears to interact with fetal and/or maternal genotypes at specific loci such as the methylenetetrahydrofolate reductase gene [3].

In the common cancers, it has become clear that there is a sub-group of affected individuals with, effectively, a Mendelian dominant family history of the same condition. For breast cancer, for example, some 5% of cases arise in women with a strong familial predisposition. Much of this predisposition is accounted for by mutations at two loci, *BRCA1* and *BRCA2*. The genetic basis for differences in susceptibility to such cancers among the rest of the population has been more difficult to dissect.

It has been suggested that such advances in our understanding of the genetic component of common, multifactorial disorders will lead to a 'paradigm shift' in health care [4]. The dissection of the genetic factors predisposing to these disorders will (as maintained by Baird), lead to the identification of individuals at increased risk of these diseases; in turn, this will lead to health benefits because 'forewarned is forearmed'. Those at high risk of these disorders will be able to alter their lifestyles so as to improve their future health, and they will be motivated to do so because of their knowledge of their personal high risk of future ill health.

An even more enthusiastic assessment of the role of genetic technology in health care has come from Weber [5], a scientist renowned for his contribution to genetic linkage analysis using highly informative microsatellite markers. In 1994, he proposed the immediate banking of DNA from all newborns and all the elderly and the sick, so that their DNA can be typed with 250+ markers from across the genome: " . . . global screening of polymorphisms enables the entire genome to be examined in one step". Apart from revealing cases of false paternity, it is not clear what application this information would have, except in very specific circumstances where the necessary genetic studies are already performed in simpler, more appropriate ways. There may be some logic to banking DNA from certain elderly or sick individuals, but what could be the point of typing so many polymorphisms in so many people when there is no concrete use to which the information would be put? The information generated would be potentially disruptive socially, while being irrelevant to health care and too crude to be useful for research. This is a case of enthusiasm gone wild.

Baird does see some potential dangers in the application of the new technology: "If our society is to take advantage of the use of genetic testing to give the opportunity for avoidance of disease, the testing will have to be acceptable to the citizens. If genetic testing becomes stigmatizing and socially handicapping (e.g. so someone cannot get insurance or employment), we will not, as a society, be able to benefit from these new approaches to disease prevention." Even Weber raises a caution about genetic discrimination. These cautions, however, are not given much emphasis. The application of genetic technologies to disease prediction and prevention (susceptibility screening and risk factor modification) is portrayed as overwhelmingly beneficial. Like Holtzman [6], I would like to present a more cautious view.

THE APPROPRIATE DISSECTION OF MULTIFACTORIAL GENETICS

The principal scientific justification for these genetic studies of multifactorial diseases is that they will lead to a better understanding of the disease processes involved. The genetic elucidation of Mendelian diseases has led to the identification of many important biological processes and components. Similar advances are likely to be achieved through the genetic dissection of susceptibility to multifactorial disease, although such work will often require more families to be studied and will be more demanding of laboratory time and computational skills than the studies of Mendelian diseases. Given that most of the conditions ranking as major public health problems are multifactorial, the importance of this work could hardly be overstated. Seven such diseases – major causes of serious chronic morbidity and premature mortality – are ischaemic heart disease, hypertension, diabetes, strokes, breast cancer, colon cancer and Alzheimer's disease.

If successful, the genetic dissection of these diseases may lead on to improved therapies, perhaps even to gene replacement therapy but more likely to other forms of rational therapy. Through understanding disease processes more fully, it may be possible to design drugs to compensate for the underlying causes of disease. Opportunities will open up in pharmacology as newly recognized steps in cellular function become potential targets for a new generation of therapies – doubtless to be portrayed as a wonder generation of *new* 'magic bullets', at least until their limitations become apparent over time. Hypertension, for example, will no longer be regarded as a single condition. It will instead be important to identify the underlying genetic predisposition that is present in the specific patient with high blood pressure to ensure that the optimal treatment, tailored to their genotype, is prescribed. For some conditions, the supplementation or replacement of defective genetic material in some tissues may even become feasible, safe and effective.

The scientific rationale for carrying out genetic research into these diseases has never included the development of tests that will identify the risk for healthy individuals that they will develop the disease. The fact that so many genes and non-genetic factors are involved in the aetiology of these common diseases means that the identification of inherited predisposition is of little use at the individual level; it will never be possible to predict those who will be affected nor to know when an individual will develop a disease if he does so at all. Such tests may nevertheless be developed for a combination of reasons: commercial 'necessity' in the biotechnology and insurance industries, professional enthusiasm among clinicians and research scientists, and the desire in all of us to apply knowledge to important questions of human health.

LIMITATIONS TO SUSCEPTIBILITY SCREENING

For a few, uncommon genetic disorders (the Mendelian diseases), predictive testing within known families seems likely to be of real benefit in identifying those at high risk of heart disease, cancer or other serious conditions, because useful therapies or health surveillance are known to improve the outcome for those at high genetic risk. For the common, multifactorial disorders, however, there may be no public health benefit from screening the general population for their genetic susceptibility.

One problem with susceptibility screening – testing individuals to specify their personal risk of developing certain diseases – relates to the type of result that could be generated. A person's susceptibility to a common disease could be presented as a lifetime risk figure and given as a fraction, a percentage, an absolute probability or an odds ratio. It could be presented instead as a relative risk compared to an 'average' member of the population. The person's age could be taken into account, and they could be given a risk figure that applied to a fixed time-frame (e.g. the next 10 years). But such tests will not be *predictive*; they will only convey information about relative risk, and will leave as much uncertainty as they are likely to resolve. Furthermore, however carefully the risk information is conveyed, very important issues will remain about how the information is understood and used by the individual.

Another set of problems with such testing of individuals is the validity of the risk figures provided. Usually, the risk figures will be derived from datasets drawn from one or a few populations, so that differences in disease frequency and random genetic variation between ethnic groups and specific populations will limit the accuracy of results. In addition, without large numbers of subjects being followed for many years in a prospective research project, there will be concerns about bias in the reference dataset, and the confidence intervals for risk information will remain wide. For many years to come, then, the accuracy of any risk estimates provided will be suspect. This will be a particular problem when information from different sources is pooled to give a combined risk estimate on the basis of several different genetic tests applied to the same individual. There will doubtless be pressure to do this, to squeeze as much information as possible from the tests, but it will be very difficult to know how to combine risk information from different tests unless all the tests used have already been studied in the same large set of research subjects. Otherwise, it will only be surmised that the different factors being studied act independently to modify the risk of disease.

CONSEQUENCES OF SUSCEPTIBILITY SCREENING FOR THE INDIVIDUAL

The principal justification for screening will be for those 'at risk' to alter their lifestyle, and so modify their risk of various diseases. Will this, however, be of any more benefit to them than simply deciding to lead a non-specifically 'healthy lifestyle'? While some evidence indicates that a positive family history of heart disease, or a test result indicating susceptibility, may lead to marginally increased compliance with health checks, this does not amount to increased long-term compliance with health-enhancing behaviours. Furthermore, while some of those identified as being at increased risk may modify their lifestyles 'appropriately' (in accordance with medical advice), others may be fatalistic or may even react in a paradoxical fashion and exacerbate their risk factors [7].

Lay understandings of health risks in relation to lifestyle are intricate, but differ from the medical understandings of those involved in health promotion [8]. This may help to explain why experience with health promotion in primary care is not encouraging, with only modest changes in risky behaviours after very considerable investment of resources directed at effecting behavioural change [9–11]. The introduction of such screening in the absence of established benefits for those found to be at high risk is unethical [12], and the focus on genetic factors may distract attention from " . . . the real challenge of the future (which) appears to be the behavioural and social issues of risk-reduction" [13].

Individuals might respond to risk information from screening tests by selecting just how to abuse their bodies, by identifying the pattern of over-consumption that their bodies can tolerate with least harm. "The tests say that I can drink a bottle of Scotch a night, but I have to watch my blood pressure and have a check for faecal occult blood every 3 months". Would such screening really promote health?

A number of problems remain, even if the high-risk subjects are placed on a package of risk-lowering measures. The ability of those at risk to influence their disease susceptibility may be limited – there may be no established intervention of proven benefit – and the compliance of at-risk individuals with the recommended behaviour pattern, medication or diet may be poor. These limitations of lifestyle modification serve to emphasize the importance of careful pre-test and post-test information and counselling when screening is provided for a number of different conditions.

PUBLIC HEALTH CONSEQUENCES

Population susceptibility screening for multifactorial disorders will be justified only once careful evaluation has demonstrated clear benefits. These benefits would have to extend not just to those identified as being at high risk but would have to include those at low or average risk too. Such proof of overall benefit to the population is most unlikely until improved methods of treatment or surveillance for complications have been established. Until that time, perhaps far distant, there is no reason to adopt the privatized screening approach to produce individualized indicators of susceptibility. Indeed, in so far as health benefits can be gained by altering lifestyles, the benefits of a healthy lifestyle accrue to all, not just to those at a personally increased risk of a specific disease [14–16]. Individualizing genetic risks will be inefficient and expensive (the corollary of its being profitable), requiring capital investment in technology and promotion, and leading to few benefits that could not be achieved by the general adoption of a healthy lifestyle without the costs of individual susceptibility testing [17].

It is even possible for the overall population to be affected adversely by screening designed to identify those at high risk. Those not identified as being at increased risk may be falsely reassured that they will not develop that particular condition, and they may then fail to modify their lifestyle in the appropriate manner. They may fail to appreciate that those at average or even at low risk of a common disease will still increase that risk if they ignore lifestyle advice, which can therefore still be of benefit to them [18]. Those identified as being at increased risk (the 'worried well') may become fatalistic, and decide that heart disease or bowel cancer is inevitable, and so they may as well enjoy their life and indulge themselves. Identification as being at increased risk could then become a self-fulfilling prophecy. This potential for paradoxical, even perverse, consequences of ostensibly rational, medically 'sensible' screening programmes must be remembered, so that evaluations of such screening are able to identify the harmful consequences of screening programmes as well as the beneficial effects.

The potential for harm being caused by susceptibility screening is especially clear in relation to predisposition to psychiatric disease. Those who are predisposed to such illnesses may well be more vulnerable to emotional distress, and being identified as susceptible could then all too easily become a self-fulfilling prophecy. The suggestion that a knowledge of susceptibility could be helpful through permitting the vulnerable individual to modify their lifestyle certainly assumes that environmental stresses could be avoided, but ignores the fact that such information could itself be a powerful stressor.

A HEALTHY LIFESTYLE?

The discussion so far begs the question of what is a healthy lifestyle. In so far as there is an answer supported by evidence, a more substantial and more equitable improvement in health could be achieved through political action to promote the health of the entire population rather than concentrating resources on defining personalized disease susceptibilities [19]. Such concerted efforts would include the promotion of a generally healthy diet (low in cholesterol, high in fresh fruit and vegetables) and of modest physical exercise (encouraging public transport and cycling), as well as the enforcement of tougher road safety measures and air pollution standards and the discouraging of the consumption of tobacco. There would need to be substantial financial weight behind both the stick and the carrot. It is not possible to make much more specific proposals about the fat or sugar content of the diet until the epidemiological evidence is clearer – the same limitation, of course, applying to advice given at a susceptibility testing clinic.

The prospect for such action in support of public health in Britain may seem remote, when it would entail large-scale intervention in the 'free market' in relation to food, tobacco and agriculture, and a restructuring of public and private transport and of town planning. These problems demand co-ordinated, collective responses; ideologically fixated, 'spinal-reflex' preferences for individualized, privatized solutions will only make matters worse [20].

RESEARCH ISSUES

Specific interventions may soon be devised to prevent the development of disease in those identified as being at high risk (e.g. of diabetes or asthma). Careful research into the risk factors and preventive measures relevant to each disease should be encouraged, which may lead to evidence that justifies trials of screening, but it will be essential for the outcomes of those identified as being at low risk to be measured as well as the outcomes of those at high risk. It will also be important for clinical research into medical outcomes to be shadowed, or paralleled, by research into the psychosocial and behavioural consequences of risk identification and risk-factor modification (for those at high *and low* risks).

It may be difficult to demonstrate the efficacy of preventive interventions, however, without first identifying those at high risk of the disease to be prevented. Only with access to such high-risk groups will it be realistic to mount trials of new potential remedies or preventive agents. This introduces another potential ethical problem: the danger of recruiting individuals into susceptibility screening projects who think this will be for their good when it is not so.

It will be necessary to ensure that screening programmes to identify those at high risk are clearly presented to all potential participants as research studies until the benefits of specific interventions have been established. If the labelling of a research-driven screening programme as a research exercise is not very clear, then subjects may be recruited and labelled as high risk only to find that there is no clear benefit. Indeed, such subjects may feel coerced to participate in subsequent trials of possible interventions, because that may seem preferable to simple passivity. This situation simply must not be allowed to arise; no susceptibility screening programme should be established in the guise of a regular service when there is no benefit to the participants, and when the real purpose is to recruit subjects into trials of possible risk-modifying interventions. The finding that those at risk of inherited cancers may volunteer to participate in research because they anticipate that this will give them access to better care [21], makes it especially important not to identify those at-risk unless those screened understand that there may be no benefit from screening.

COMMERCIAL INTERESTS AND THE DISTORTION OF SCIENCE AND MEDICINE

Human genome research is being conducted in a commercial climate, and the prospect of the application of molecular genetic technology to the general (healthy) population in genetic susceptibility screening programmes is one of the major reasons for this. While susceptibility screening may be bad science, it is likely to be excellent business. Screening tests applicable to the general population will hold out promise of enormous profits for those corporations that can develop and patent tests and techniques ahead of their competitors. The venture capital that has been invested in biotechnology corporations over the past few years, will be required to produce a substantial rate of return, and the rewards from rational therapy and gene therapy still appear too remote to satisfy investors – or even perhaps the scientists themselves.

Control of the technology required for genetic susceptibility screening is largely in private, corporate hands. This will have important consequences for the technical development and the promotion and use of such tests. The expense of the tests will ensure that they are introduced into private health care systems ahead of proper evaluation, and ahead of their introduction into state insurance schemes like the National Health Service (NHS) in the UK. This will be a blessing in disguise if the tests do not yield the promised benefits, but if there are health benefits they will be introduced inequitably, because access will be restricted to those who can pay (whether directly or through insurance companies).

One difficulty likely to result from the commercial control of susceptibility screening will be its active promotion. We know from pilot studies of carrier screening for cystic fibrosis that the mode of offering a genetic carrier test is much the most important factor influencing uptake of the test. We know from life in the late 20th century that this is true for almost all consumer goods as well as genetic tests: presentation and packaging are everything. Those who respond to the marketing and choose to participate in such screening will be helped to feel rational, autonomous and indeed fortunate – in control of their biological destiny. The fostering of this illusion by advertising agencies will be carefully constructed but inconspicuous. Those who do not choose, or cannot afford, to participate in such screening may be portrayed as irresponsible and feckless. This implicitly promotes the notion that genetic endowment and chosen lifestyle, together determine one's future health, while the importance of material circumstances (especially poverty) in creating ill health will be glossed over [22–24].

In a commercial context, it may not be in the financial interest of the testing agency to provide full pre-test information, because it would be expensive and could deter significant numbers of potential customers if the limitations of lifestyle modification were explained.

One aspect of susceptibility screening that is likely to be omitted from the promotional literature and videos is the psychological burden of being identified as susceptible. Experience with blood pressure and serum cholesterol screening programmes [25,26] indicates that problems are likely. How will these counselling issues be addressed in a commercial environment? There is a danger that pre-test counselling could be confused with sales talk.

The identification of large numbers of individuals (more 'at-risk' individuals will be identified than there would ever be patients with the relevant diseases) may also have enormous repercussions for those trying to provide a national health service with already inadequate resources. Support through the distress produced by positive results is unlikely to be provided by the gene testing corporations, and much of it will land at the door of the national health service or voluntary agencies, as happens already for families distressed by unwelcome results from commercial maternal serum screening for Down syndrome. It is the national health service that is likely to pick up the cost of any drugs or diet recommended because of the corporations' test results, and it is staff in the national health service that will provide most of the other services that are sought as a result of testing – the explanations, the genetic testing of other family members, the emotional support, the lifestyle advice. . . . The facilities of the national health service will be burdened by commercial genetic testing programmes, which will therefore not be bearing the full social cost of their activities. The under-resourced national health service will be subsidizing private capital's adventures with tests of unproven worth. Similar considerations will apply in other health care systems where individuals can obtain genetic screening outside their health insurance schemes.

SUSCEPTIBILITY SCREENING AND GENETIC PRIVACY

There will be important, adverse consequences for individuals identified as being at increased risk of disease, which may not be pointed out to them in advance of testing. Screening may make it difficult for them to obtain insurance or employment. Companies will seek employees who are likely to remain physically and mentally healthy until aged 60–65 years, and who are of low susceptibility to the relevant occupational diseases. Insurance companies will feel compelled to modify their premiums or the cover they provide, or to deny insurance altogether, in the light of susceptibility test results, especially if such tests are readily available to their clients or competitors (see Chapter 4). If insurance companies have access to the results of such tests on their policy-holders or applicants, there may also be concern that the results will be made available through pooling of risk information across the insurance industry, or through sharing risk information with employers.

These tests will not be legally compulsory, but they may become effectively obligatory for substantial sections of society – those who choose to take out life or health insurance, or who seek managerial or professional employment, or employment in a setting with specific occupational health hazards. In the 'free market' of commerce, such tests may be portrayed as voluntary, but failure to 'volunteer' may lead to exclusion from full participation in society.

The use of DNA technology to identify those with better or worse health prospects, a genetic sorting into sheep and goats, could create social inequalities, or more likely be used to reinforce existing inequalities in access to health care, education, careers and the material goods of our age. This notion of a genetic underclass resonates with the echoes of eugenics from earlier this century. The contemporary version would be more soundly based in scientific terms, and the discrimination would be market-led rather than state-enforced, but the consequences could be comparable. The distribution of most risks in society is heavily influenced by wealth and power [27]; genetic risk would be different in that it could be used to influence access to wealth and power, and hence exposure to other, socially regulated risks.

These reasonable concerns amount to a strong additional reason (if any additional justification were needed) to favour national health care schemes that ensure universal access on equitable financial terms. Whether a country's scheme is financed through compulsory insurance or taxation is immaterial; it is only if the whole population is covered that individualized health risks can be ignored in calculating the charges borne by each citizen.

SUSCEPTIBILITY TESTING OF CHILDREN

An additional concern would be that children may be tested, whose parents are naturally concerned about the future well-being of their child. From excessive zeal, such parents could launch their child on an unbalanced, unproven or even frankly harmful programme of diet or exercise, resulting in paradoxical long-term physical and emotional harm to the child [28,29]. The identification of children as being different from their peers, as genetically flawed and potentially diseased, may have profound consequences for the socialization of these children – and there could be serious implications for their future personal happiness and social behaviour as adults. Stigmatization at an early age could be very damaging.

Attention was drawn to the emotional consequences of inappropriately labelling children as sick 27 years ago in the context of benign cardiac murmurs [30]. There have been considerable problems arising out of cholesterol screening in children in the USA, ranging from frank malnutrition [31] to poor parental compliance with a cholesterol screening programme [32]. In a follow-up study of newborn screening for α_1-antitrypsin deficiency, fathers of children at risk of serious lung disease actually smoked more heavily than did the fathers of control children [33]. For cholesterol, despite much research, the case for population screening of children has not been made, although there is a strong case for testing serum cholesterol in children from families with inherited hyperlipidaemias [34]. Families may focus excessive attention on the child at risk, or may deny the risk and fail to meet the child's real needs.

The experience of predictive testing for Huntington's disease in The Netherlands also lends considerable support to a policy of extreme caution. Those who became aware in adulthood of their risk of future disease seem to have coped better than those who learnt of their risk status in adolescence [35].

In addition to these concerns, there are the needs of other sibs to be considered, the unaffected ones, and the complex process of family adaptation to test results that has to recognize feelings of guilt and resentment in those *not* at risk. When some children in a family are shown to be at increased risk and others not, the potential complexity of the interpersonal consequences is enormous and could well be damaging.

The genetic testing of children is considered in a broader context in Chapter 2.

MEDICALIZATION AND GENETICIZATION

The final disadvantage of genetic susceptibility testing to be discussed here is that it promotes medicalization and geneticization. The medicalization of life will be promoted because of the large number of individuals who will be identified as being at increased risk of something. Some of these individuals will then seek medical advice and alter their lifestyles, while others will ignore the results but worry. Both groups will suffer from this invasion of technical medicine into their lives.

Geneticization will also be promoted; this is the explanation of differences between individuals and groups in purely genetic terms, and the consequent overemphasis of genetic factors in planning the provision of health care. Such a focus could lead to expensive, high-technology 'solutions' to problems that might be better tackled by social or environmental means – but which would not lead to profits for the gene corporations or for the scientists who work in the blurred zone between academic research and the commercial applications of biotechnology.

Abby Lippman has drawn attention to the problems of geneticization: " . . . it is not immediately apparent that we need to know where genes are located to improve the overall health of a population or to promote the well-being of individuals, to understand illness, or to relieve suffering. We could, in fact, 'map' the environment for sources of 'susceptibility' instead of mapping the genome" [36].

Both types of mapping are necessary for a balanced account of human disease [37], which in turn is needed for the effective and equitable provision of health care. Mapping human disease to genes, and unravelling the genetic contribution to common disorders, has many potential benefits; the application of this new knowledge to individualized screening of health risks, however, may not be one of them.

The suggestion that health in the Western world could be improved by collective measures attending to social and environmental factors may be viewed with suspicion by those who have integrated Baird's genetic paradigm of modern medicine without retaining a proper understanding of the interaction between genotype and environment, including the social environment [38]. The links between health and poverty and the other material conditions of life must not be forgotten. Geneticization promotes a blinkered view of the world that exaggerates personal responsibility for health, denigrates the collective solutions to health problems that may be the only hope for those with few resources, and favours personal and corporate profits over the collective provision of equitable health care. In these ways, geneticization may seriously impair the attempts of medicine and of society to promote health for all.

CONCLUSIONS

Genetic susceptibility screening of those at population risk of serious diseases should only be introduced cautiously and with safeguards. It should be considered when careful clinical and psychosocial evaluations of pilot programmes have demonstrated direct benefits from pre-symptomatic intervention in those who can be identified as being at increased risk of the disease in question. The evaluation of such programmes should examine the outcomes for those who are given reassuring results (normal or low risk), as well as of those at increased risk of disease.

Programmes of population screening for genetic susceptibility to disease should not be introduced, sponsored, managed or promoted by commercial corporations but, where of proven benefit, should be accessible to all through a national health service. Pre-test information and explanation should be provided, and post-result follow-up, support and counselling should be available for those at increased risk. The provision of testing without full provision for counselling and support should not be permitted.

Genetic testing of those in specific high-risk groups, or who believe themselves to be at high risk, such as those with a strong family history of cancer or heart disease, may be appropriate, even in the absence of medical interventions of proven worth. The issues to be considered are different.

Social and political decisions that ameliorate poverty and that alter lifestyles appropriately may enhance the health of the general population more effectively and more cheaply than individualized susceptibility screening to identify those at increased risk, although such policies may be unpopular with some because they will generate less profit for the biotechnology and pharmaceutical industries.

REFERENCES

1. Davies JL, Kawaguchi Y, Bennett ST *et al.* (1994) A genome-wide search for human type 1 diabetes susceptibility genes. *Nature*, **371**: 130–135.
2. Hashimoto L, Habita C, Beressi JP *et al.* (1994) Genetic mapping of a susceptibility locus for insulin-dependent diabetes mellitus on chromosome 11q. *Nature*, **371**: 161–164.
3. Mills JL, McPartlin JM, Kirke PN, Lee YJ, Conley MR, Weir DG, Scott JM (1995) Homocysteine metabolism in pregnancies complicated by neural-tube defects. *Lancet*, **345**: 149–151.
4. Baird PA (1990) Genetics and health care. *Persp. Biol. Med.* **33(2)**: 203–213.
5. Weber JL (1994) Know thy genome. *Nature Genetics*, **7**: 343–344.
6. Holtzmann NA (1992) The diffusion of new genetic tests for predicting disease. *FASEB J.* **6**: 2806–2812.
7. Davison C, Frankel S, Davey Smith G (1989) Inheriting heart trouble: the relevance of common-sense ideas to preventive measures. *Health Ed. Res.* **4**: 329–340.
8. Davison C, Frankel S, Davey Smith G (1992) The limits of lifestyle: re-assessing 'fatalism' in the popular culture of illness prevention. *Soc. Sci. Med.* **34**: 675–685.
9. Family Heart Study Group (1994) Randomised controlled trial evaluating cardiovascular screening and intervention in general practice: principal results of British family heart study. *Br. Med. J.* **308**: 313–320.
10. Imperial Cancer Research Fund OXCHECK Study Group (1994) Effectiveness of health checks conducted by nurses in primary care: results of the OXCHECK study after one year. *Br. Med. J.* **308**: 308–312.
11. Stott NCH (1994) Screening for cardiovascular risk in general practice (editorial) *Br. Med. J.* **308**: 285–286.
12. McCormick J (1994) Health promotion: the ethical dimension. *Lancet*, **344**: 390–391.
13. Williams RR (1988) Nature, nurture and family predisposition. *N. Engl. J. Med.* **318**: 769–770.
14. Chen Z, Peto R, Collins R *et al.* (1991) Serum cholesterol concentration and coronary heart disease in population with low cholesterol concentrations. *Br. Med. J.* **303**: 276–282.
15. Law MR, Thompson SG, Wald NJ (1994) Assessing possible hazards of reducing serum cholesterol. *Br. Med. J.* **308**: 373–379.
16. Law MR, Wald NJ, Thompson SG (1994) By how much and how quickly does reduction in serum cholesterol concentration lower risk of ischaemic heart disease? *Br. Med. J.* **308**: 367–373.

17. Wald NJ, Law M, Watt HC *et al.* (1994) Apolipoproteins and ischaemic heart disease: implications for screening. *Lancet*, **343**: 75–79.
18. Kinlay S, Heller RF (1990) Effectiveness and hazards of case finding for a high cholesterol concentration. *Br. Med. J.* **300**: 1545–1547.
19. Mant D (1994) Prevention. *Lancet*, **344**: 1343–1346.
20. Laver M (1981) *The Politics of Private Desires: the Guide to the Politics of Rational Choice*. Penguin, Harmondsworth, UK.
21. Green J, Murton F, Statham H (1993) Psychosocial issues raised by a familial ovarian cancer register. *J. Med. Genet.* **30**: 575–579.
22. Townsend P, Davidson N, Whitehead M (eds) (1988) *Inequalities in Health: the Black Report and the Health Divide*, 2nd Edn. Penguin Books, Harmondsworth, UK.
23. Townsend P (1990) Individual or social responsibility for premature death? Current controversies in the British debate about health. *Int. J. Hlth Serv.* **20**: 373–392.
24. Phillimore P, Beattie A, Townsend P (1994) Widening inequality of health in northern England, 1981–1991. *Br. Med. J.* **308**: 1125–1128.
25. Lefebvre RC, Hursey KG, Carleton RA (1988) Labeling of participants in high blood pressure screening programs. Implications for blood cholesterol screenings. *Arch. Intern. Med.* **148**: 1993–1997.
26. Brett AS (1991) Psychologic effects of the diagnosis and treatment of hypercholesterolaemia: lessons from case studies. *Am. J. Med.* **91**: 642–647.
27. Beck U (1992) *Risk Society*. Sage, London.
28. Newman RT, Browner WS, Hulley SB (1990) The case against childhood cholesterol screening. *J. Am. Med. Assoc.* **264**: 3003–3005.
29. Harper PS, Clarke A (1990) Should we test children for 'adult' genetic diseases? *Lancet*, **335**: 1205–1206.
30. Bergman AB, Stamm SJ (1967) The morbidity of cardiac nondisease in schoolchildren. *N. Engl. J. Med.* **276**: 1008–1013.
31. Lifshitz F, Moses N (1989) Growth failure. A complication of dietary treatment of hypercholesterolaemia. *Am. J. Dis. Child.* **143**: 537–542.
32. Bachman RP, Schoen EJ, Stembridge A *et al.* (1993) Compliance with childhood cholesterol screening among members of a prepaid health plan. *Am. J. Dis. Child.* **147**: 382–385.
33. Gustavson K-H (1989) The prevention and management of autosomal recessive conditions. Main example: alpha$_1$-antitrypsin deficiency. *Clin. Genet.* **36**: 327–332.
34. Lloyd JK (1991) Cholesterol: should we screen all children or change the diet of all children? *Acta Paediatr. Scand.* **373** (Suppl.): S66–72.
35. van der Steenstraten IM, Tibben A, Roos RAC *et al.* (1994) Predictive testing for Huntington disease: nonparticipants compared with participants in the Dutch program. *Am. J. Hum. Genet.* **55**: 618–625.
36. Lippman A (1993) Worrying – and worrying about – the geneticization of reproduction and health. In: *Misconceptions*, Vol. 1, pp. 39–65 (eds G. Basen, M. Eichler, A. Lippman). Voyageur Publishing, Quebec.
37. Weatherall D (1992) The role of nature and nurture in common diseases. Garrod's legacy. The Harveian Oration of 1992. Royal College of Physicians, London.
38. Ramsey M (1994) Genetic reductionism and medical genetic practice. In: *Genetic Counselling. Practice and Principles*, pp. 241–260 (ed. A. Clarke). Routledge, London and New York.

Chapter

8

NEWBORN SCREENING

A.J. Clarke

INTRODUCTION

The development of newborn screening is one of the great success stories of twentieth-century preventive medicine, and a recapitulation of this tale covers several of the most exciting developments in genetics and biochemistry. The story began in 1902 when Archibald Garrod categorized the familial condition alkaptonuria as an inborn error of metabolism. A much more serious condition, phenylketonuria (PKU), was recognized as another such inborn error of human metabolism by Folling in 1934. Every student of genetics learns about the work of Beadle and Tatum, who studied mutations that affect amino acid metabolism in the fungus *Neurospora*. They demonstrated the accumulation of intermediate compounds in a metabolic pathway caused by a lack of functional enzyme. This work in a lowly fungus led them to the recognition of the crucial relationship between genes and enzymes: put simply, one gene (usually) corresponds to one enzyme.

This understanding of inherited metabolic diseases as blocks in a metabolic pathway also led to the development of a strategy for the treatment of inborn errors, which in turn led to the development of a dietary treatment for PKU. The outlook for the second affected child in a family was greatly improved when the condition was recognized early and treatment was started promptly, and this provoked the search for a practical screening test to detect the condition in seemingly healthy (presymptomatic) newborn infants. Guthrie's bacterial growth assay for elevated blood levels of phenylalanine, utilizing a dried blood spot on filter paper, became the archetypal population screening programme for genetic disease; by screening the whole population of apparently healthy newborn infants, the presymptomatic, affected infants can be recognized so that effective dietary treatment can be instituted.

PKU affects about one child in every 10 000 in Britain, causing severe intellectual impairment, but its ill-effects can largely be prevented if an affected child adheres to a suitable low-protein diet from soon after birth [1]. Although the laboratory methods have since changed, this approach to the prevention of profound intellectual disability through the early diagnosis of affected infants has continued. The same logistic framework – the collection of dried blood spots from infants in the first week or so of life – has been utilized to incorporate an assay for thyroid hormones. This identifies the one in 4000 infants who are born with congenital hypothyroidism, most of whom will benefit greatly from treatment with thyroid hormone [2].

Newborn screening has saved many thousands of infants from these two severe disorders, and has thereby accumulated a large store of goodwill and of ethical credit in favour of genetic screening programmes. Indeed, newborn screening for PKU and hypothyroidism may be regarded as 'the acceptable face' of genetic screening. There is no doubt that these programmes, and newborn screening for sickle cell disease where it is sufficiently frequent, meet the established World Health Organisation guidelines for the adoption of screening programmes [3]: these diseases are important health problems, the testing process is socially acceptable and is technically and financially feasible, and the treatments are known to improve the outlook for affected individuals. Even in this type of genetic screening programme, however, there can be problems.

INFORMED PARENTAL CONSENT

The first issue we will address relates to the nature of the parental consent to newborn screening. As emphasized by Clayton [4], newborn screening programmes are unusual in that they are provided by the state and are often the only medical test offered to, or even required from, all its citizens. What justifies this interest of the state in newborn screening? Even in developed countries, where newborn screening is nearly universal in its coverage of the population, there is a great diversity of systems and practices in operation. In some countries newborn screening is mandatory, in others it is universal because it has been routinized and is scarcely ever questioned or challenged by parents, and in others it is offered on the basis of informed parental consent. Policies vary even within countries – between states in the USA and even between districts within Wales.

Because newborn screening for PKU is carried out for the sake of the individual infant, in his/her own best interests, it could be argued that screening should be universal and mandatory. This would mark out newborn screening as being very different from other genetic screening programmes, where there is a general consensus in developed countries that participation should be voluntary and that decisions should be made on the basis of informed consent.

There have been few empirical studies that have addressed the issue of informed parental consent for newborn screening. It is clear from the survey of Statham *et al.* [5], that parental knowledge about newborn screening is often rudimentary. It is also clear that the level of maternal knowledge about newborn screening can be greatly improved if a policy decision is taken to offer newborn screening on the basis of informed parental consent rather than imposing it, and such a policy can be inexpensive to implement [6,7]. Changing the system of newborn screening to one based on informed consent need not be accompanied by an increase in the proportion of infants who are not screened [6], but is the extra information about testing at all useful to parents? When we turn to consider screening for disorders where the benefits to the affected child are less, participation on the basis of informed consent is clearly vital – but does it matter when the screening is carried out on behalf of the child?

My answer is that screening on the basis of informed consent will be preferable because of the ethos that this will foster in the health care system. Such a policy will promote the concept of partnership between parents and professionals in providing health care for children. It will also make it more likely that other choices about newborn screening that may be introduced in the future will be introduced as genuine choices that parents discuss and consider before making a decision, in keeping with the spirit of the (British) Cumberledge Report 'Changing Childbirth' [8]. Both the health professionals involved and the parents, will then be used to a system in which the voice of each individual counts. This will be especially important when the benefits to the individual child from newborn screening for certain additional disorders are less definite than with PKU.

ACCESS TO TESTING AND TREATMENT

Given that a programme of newborn screening is of benefit to the child, it is clearly important that access to the screening is as close to universal as possible. When there are effective treatments available for affected children, it is vital that those identified by screening have ready access to them. Where the state imposes the newborn screening programme (as in parts of the USA) it is especially important that financial or other barriers to treatment do not prevent some children – the children of the poor – from obtaining the full benefit of an early diagnosis.

In the USA (where a second newborn screening test is generally recommended because the first test is carried out too soon after birth for all infants affected to be identified) there are even more problems. While uptake of the first test is very high, the proportion of infants who receive the second test varies greatly and social factors are important influences on this. Maternal ethnicity, age, and years of education and the number of prenatal care visits are important variables that influence this [9]. Given the context of a private health care system largely funded by insurance, there are dangers that a child identified as affected by the (in many states) mandatory newborn screening programme, will be uninsurable; in that case, the family may have to meet the costs of treatment without state or insurance support [10]. This is clearly a serious failing, and serves as a warning to anyone who wishes to see the British system move further away from the National Health Service to a private insurance-based scheme.

Even in Britain, where cost considerations should not distort patterns of care along the lines of social class because of the relatively equitable access to National Health Service provision, access to newborn screening is influenced by social factors. Screening for congenital hypothyroidism was studied in 10 English districts over two years. The proportion of infants who were not screened was 2.6% overall, with the figure ranging from 0 to 3.8% in different districts. Of the 981 infants not screened, there were only nine cases in which parents had refused screening; the other failures were the result of logistic and organizational deficiencies, and follow-up of these failures was incomplete [11]. In a study from south London, there was substantial variation in coverage of newborn infants by ethnic group and by district: an infant of African (not Afro-Caribbean) ethnic origin had a three-fold relative risk of not being screened compared to a white infant; infants from one district had a two-fold risk relative to infants from the other district studied, after allowance had been made for ethnicity [12]. These studies show that newborn screening programmes need to be constantly monitored to ensure that access to the benefits of screening is not denied to children because of factors related to social class or ethnicity; they also show that regular audits are required to ensure that newborn screening programmes operate efficiently.

PSYCHOLOGICAL RESPONSES TO NEWBORN SCREENING

There has been little work carried out to assess how families respond to the information that their child has a problem identified by newborn screening for PKU or hypothyroidism; the focus of evaluations has been on the clear health benefits of early diagnosis. Psychological responses to newborn screening have been examined in relation to newborn screening for disorders where the health benefits of early diagnosis are less clear-cut.

Newborn screening for cystic fibrosis (CF) has been made available in a number of countries and using a range of different methods. Before molecular genetic testing became feasible, the most widely adopted method was a two-stage procedure measuring the infant's serum immuno-reactive trypsin (IRT) level, which is elevated in many infants with CF. In those infants in whom the IRT was elevated on the first round of testing, a second blood spot was taken; if the IRT was still elevated, a sweat test was performed to identify those who were actually affected. This process entailed subjecting a large number of infants to further investigations, and hence causing distress to their families. In order to identify 85–90% of affected infants, a second blood spot had to be taken from at least three or four infants per 1000 [13]; even in a population with one affected infant in 2500 births, at least 90% of those from whom a repeat sample was obtained would turn out to be unaffected [14,15]. This approach led to a great deal of anxiety in a large number of families and, in one study, more than one-third of the parents of the unaffected children suffered lingering concerns about their child's health [16]. It is particularly important to note that these concerns were greater in parents who reported that they had not received sufficient information about the screening process and its significance.

The case for newborn screening to identify infants affected by CF has now been strengthened. Not only is there persuasive evidence – although not absolute proof – that affected infants remain healthier if they are diagnosed by screening [17–21], but the advent of molecular genetic diagnostics means that CF gene mutations can be sought in those infants with a raised IRT on the first round of testing, without any need to contact the family for a further sample. Families will only need to be contacted when a child with a high IRT also has at least one CF gene mutation [22]. Some of the affected infants will be identified as having two disease-causing CF gene mutations, while other infants will be found to have a high IRT and just a single CF gene mutation – these will either be healthy carriers or they will be affected infants in whom only one of the two mutations is recognized by the molecular tests. A sweat test will be required for those with a single mutation identified to distinguish between these two possibilities. This methodology has the great advantage – compared to the earlier method – that at least 50% of infants identified by the screening process will be affected, so the number of false-positive test results is much reduced; all the other identified infants will be carriers. Furthermore, the cost of diagnosing cases through screening is less than the cost of making the diagnosis in a more traditional fashion [23], and families whose affected child is identified through newborn screening are reported as finding this emotionally helpful [24]. As improved methods are developed for identifying mutations at the CF gene, it may become possible to identify almost all infants affected with CF among those with elevated IRT simply by molecular genetic testing. This could then be used to ignore the likely carrier infants with a single identified mutation who would be most unlikely to be affected. Further research into the impact of CF newborn screening on families will be required to help decide whether it is better to notify every family in which a mutation is found, even if the child is likely to be only a carrier, or only those families in which the child is shown by the gene testing to be affected.

When newborn screening does not lead to very clear health benefits for the affected infants, the impact of the screening process on the families of identified infants becomes a major factor in deciding whether the newborn screening programme is worthwhile. Screening for CF probably does lead to sufficient benefits to justify screening, but the effect is not great and it will be most important for regions that introduce CF newborn screening to monitor its impact on families to ensure that any emotional disruption caused by the programme is kept to a minimum. In our experience, it is necessary to adopt a careful set of procedures for liaison between the screening centre, the primary care team and the local paediatrician if problems are to be avoided. It is most important that the members of the primary care team who approach the family have been briefed about the significance of the test result for the child, and that the clinical information they have about CF is correct.

Another disorder for which newborn screening has been attempted is α_1-antitrypsin deficiency (α_1AT). This condition leads on to serious liver disease in a small proportion of affected infants, and to chronic obstructive airways disease in adult life (emphysema), in a larger proportion, especially if the affected individual smokes tobacco or suffers occupational exposure to dust. The principal justification of newborn screening for α_1AT, in fact, would be to protect the children from exposure to tobacco smoke and other potential lung irritants. Screening for α_1AT was introduced in Sweden in 1972 and was continued for nearly 2 years. Many parents of affected infants became anxious, worried and fearful, and these responses were often strong and long-lasting. The programme was discontinued because of this. It was also found that the parents of affected children (especially the fathers), smoked more than other fathers, and so the one potential health benefit to the affected children that could have materialized, in fact failed to do so [25,26]. There are lessons to be learnt here in relation to susceptibility screening (see Chapter 7), as well as other newborn screening programmes.

NEWBORN SCREENING FOR DUCHENNE MUSCULAR DYSTROPHY

There is a wide range of other disorders for which newborn screening is possible, but which fail to meet the WHO criteria for a screening programme [3]. These conditions include galactosaemia, congenital adrenal hyperplasia, biotinidase deficiency and medium chain acyl-CoA dehydrogenase deficiency; despite strong advocacy by enthusiasts, screening for these conditions has not become widespread because there is insufficient evidence that the natural history of the disorder is influenced by screening. For example, the disorder may often have presented and been diagnosed clinically before a screening diagnosis could be made, or the response to treatment may be uncertain or disappointing. The development of tandem mass spectrometry as a screening tool for metabolic disease will very considerably broaden the scope of screening for inborn errors, but whether or not this will lead on to improved long-term outcomes is not known. This chapter is not the place in which to review the arguments for and against newborn screening for a host of these rare and complex metabolic disorders.

What it is appropriate to broach here is the set of issues relating to newborn screening for Duchenne muscular dystrophy (DMD). This is a serious, X-linked disease that presents in early childhood with signs of muscle weakness especially in the pelvis and legs; it progresses inexorably so that an affected boy typically becomes unable to walk at 8–12 years and is likely to die at 15–25 years. While the affected boys can be helped to cope with their condition, there is no cure; DMD cannot be put right. Screening for DMD therefore identifies infants with a fatal and incurable disease. So how could newborn screening for this condition even be contemplated?

Newborn screening for DMD has been technically feasible since 1975 [27], and has been considered by health professionals because of two major problems relating to the traditional diagnostic process that have commonly been experienced by families with an affected boy. First, many families have experienced delays in having the diagnosis of DMD made in their son, and the protracted diagnostic process – often a diagnostic Odyssey – has led to bitterness, anger, distress and dissatisfaction in the parents [28,29]. Secondly, the parents of an affected boy may have further children before their first affected child has been diagnosed; if the mother is a carrier of DMD then she may have one or more further affected sons before the first affected boy is diagnosed [30].

The suggestion that newborn screening for DMD might be helpful, to avoid diagnostic delays and to permit families to seek genetic counselling before embarking upon another pregnancy, dates back many years. There was concern, however, that newborn screening for DMD might generate greater distress than a traditional diagnosis, and it might interfere with the developing relationship between the child and his parents [31,32]. It was therefore decided to establish and evaluate a programme for screening boys who were not walking independently by 18 months, taking a blood spot to assay the serum creatine kinase. It was hoped that this would reduce the diagnostic delays without causing as much distress as newborn screening.

Screening boys who are not walking by 18 months of age would never be able to identify all boys with DMD because about 50% of them walk before 18 months [31], and because any screening programme is likely to have some logistic problems and some families who choose to opt out of testing. The experience gained in Wales demonstrated the great practical and logistic barriers to establishing an effective screening programme for boys who were not walking independently at the age of 18 months. Fewer than 80% of eligible boys had the developmental assessment as scheduled for 18 months, and of those who were assessed and were not walking only 61% had a blood spot taken. Given that at least 50% of boys with DMD are walking by 18 months, the prospects for this as a screening test for DMD are not good: the programme in Wales succeeded in screening fewer than 50% of eligible boys and identified only two of the seven cases of DMD in the cohort eligible for screening [33,34].

This experience led us to reassess the case for newborn screening for DMD. A theoretical consideration of the principles was not able to resolve the dilemma: many parents who experience a later, traditional diagnosis, say that they would have preferred an early diagnosis [28], but that may not be how families feel who experience the earlier diagnosis. This is an issue that could not be resolved simply by *a priori* ethical reflection, but is one that requires experience to inform that reflection.

An attitude survey conducted among the mothers of newborn infants in Cardiff suggested that the majority (94%) would accept newborn screening for DMD [35]. There was also no empirical evidence that an early diagnosis would cause more distress than a traditional diagnosis. Any introduction of newborn screening for such a disease, however, would clearly require safeguards. We introduced an opt-in scheme, on the basis of informed parental consent, and incorporated an ongoing social evaluation of the programme and its impact on the identified families [36,37]. Furthermore, the social evaluation was used reflexively to develop and refine a protocol for handling the disclosure of the test results to the parents of possibly affected boys [38]. One strength of this protocol is that the disclosure of the medical concern about the child can be planned strategically to ensure adequate support for the family. This led to more satisfaction with the diagnostic process being expressed by the parents of boys diagnosed at neonatal screening than by the parents of boys diagnosed after a symptomatic presentation.

Since the programme began in 1990, we have demonstrated that newborn screening for DMD is technically feasible, is acceptable to most families and can operate on the basis of informed parental consent. Most families identified through newborn screening have expressed support for the programme and are pleased to have known about the condition from an early stage. The safeguards built into our programme, however, mean that any simple adoption of newborn screening for DMD in another region, as a straightforward biochemical assay introduced without attention to these same safeguards, could cause serious problems. Our protocol for handling positive cases entails a carefully co-ordinated approach to each identified family through their primary health care team and the local paediatrician [38]. Support for the primary care team from the screening centre helps to ensure continuing support for the family. The protocol we use also emphasizes the importance of parents retaining control over the diagnostic process after the initial disclosure of a potential problem; as a result, several families in Wales have decided not to proceed to a muscle biopsy to obtain diagnostic certainty, preferring instead to retain hope that their son may have the milder, Becker muscular dystrophy. Finally, the newborn screening team have worked hard to educate community midwives and other health professionals around Wales about this programme. The educational input to the midwives has helped to establish the ethos of the programme as one fostering parental choice at all stages and the very best possible care for the affected boys. To introduce screening for DMD without such a framework of support for families and for the primary care teams and without on-going evaluation of its impact on families, as yet another routine test to be added to the Guthrie card tests, would be to court disaster.

Although the early fears about newborn screening for DMD have not materialized (that it would devastate large numbers of families more thoroughly than the disease would have done by itself) there are still two unanswered questions about it that have delayed any move to introduce it as a regular part of health service provision. First, we have needed to continue our follow-up until more of the affected boys have developed frank problems from the disease; do families change their minds about newborn screening once they have experienced the full impact of the disease?

Secondly, is there an adequate level of informed consent obtained at entry into the screening programme (at the point of taking the heelprick blood sample)? Two families (out of more than 20) have expressed regrets about their participation in the screening. Could it be that the 94% uptake rate of screening for DMD indicates that the testing has become routinized, and that the quality of the informed consent obtained at entry to the screening programme may not be sufficiently stringent? One way to achieve a more appropriate uptake rate may be to select for screening the infants of parents who are motivated to play an active part in the DMD screening test, perhaps by posting the blood spot for the DMD test to the laboratory. To suggest that a lower uptake rate for a screening test would be preferable, that we should set a threshold of motivation so that infants are not screened unless their parents actively choose it, is certainly unusual but is perfectly appropriate in the context of an untreatable disease. It remains to be seen whether this policy will select out of the screening programme those families who would prefer not to know if their son has DMD.

CONCLUSIONS

Recent developments in technology – in chemical analysis as well as genetics – will make it feasible to screen for an increasingly wide range of genetic and metabolic diseases. Where an early diagnosis is known to be of benefit to the child, there are few problems. Where screening does not lead to proven health benefit to the child, however, it may still be reasonable to offer screening if the disease has its onset in childhood and if the child's family find it helpful to have an early diagnosis. In such circumstances, it will be important to ensure that parents are able to make an informed decision about newborn screening; such tests should not happen by default but only if the parents choose them. Further research is required into the advantages and the disadvantages of screening infants for untreatable disorders, examining both the process and the outcomes of screening.

REFERENCES

1. Walter J (1995) Late effects of phenylketonuria. *Arch. Dis. Child.* **73**: 485–486.
2. Tillotson SL, Fuggle PW, Smith I, Ades AE, Grant DB (1994) Relation between biochemical severity and intelligence in early treated congenital hypothyroidism: a threshold effect. *Br. Med. J.* **309**: 440–445.
3. Wilson JMG, Jungner G (1968) *Principles and Practices of Screening for Disease.* World Health Organisation, Geneva.

4. Clayton EW (1992) Issues in state newborn screening programs. *Pediatrics*, **90**: 641–646.
5. Statham H, Green J, Snowdon C (1993) Mothers' consent to screening newborn babies for disease. *Br. Med. J.* **306**: 858–859.
6. Faden R, Chwalow AJ, Holtzman NA, Horn SD (1982) A survey to evaluate parental consent as public policy for neonatal screening. *Am. J. Publ. Hlth*, **72**: 1347–1352.
7. Holtzman NA, Faden R, Chwalow AJ, Horn SD (1983) Effect of informed parental consent on mothers' knowledge of newborn screening. *Pediatrics*, **72**: 807–812.
8. Cumberledge J (Chairman, Expert Maternity Group) (1993) *Changing Childbirth. Part 1: Report of the Expert Maternity Group, Department of Health.* HMSO, London.
9. Doyle DL, Sanderson M, Bentvelzen J, Fineman RM (1995) Factors which influence the rate of receiving a routine second newborn screening test in Washington state. *Am. J. Med. Genet.* **59**: 417–420.
10. Billings PR, Kohn MA, de Cuevas M, Beckwith J, Alper JS, Natowicz MR (1992) Discrimination as a consequence of genetic testing. *Am. J. Hum. Genet.* **50**: 476–482.
11. Pharoah POD, Madden MP (1992) Audit of screening for congenital hypothyroidism. *Arch. Dis. Child* **67**: 1073–1076.
12. Streetly A, Grant C, Bickler G, Eldridge P, Bird S, Griffiths W (1994) Variation in coverage by ethnic group of neonatal (Guthrie) screening programme in south London. *Br. Med. J.* **309**: 372–374.
13. Ryley HC, Deam SM, Williams J, Alfaham M, Weller PH, Goodchild MC, Carter RA, Bradley D, Dodge JA (1988) Neonatal screening for cystic fibrosis in Wales and the West Midlands: 1. Evaluation of immunoreactive trypsin test. *J. Clin. Pathol.* **41**: 726–729.
14. Hammond KB, Abman SH, Sokol RJ, Accurso FJ (1991) Efficacy of statewide neonatal screening for cystic fibrosis by assay of trypsinogen concentrations. *N. Engl. J. Med.* **325**: 769–774.
15. Holtzman NA (1991) What drives neonatal screening programs? *N. Engl. J. Med.* **325**: 802–804.
16. Sorenson JR, Levy HL, Mangione TW, Sepe SJ (1984) Parental response to repeat testing of infants with 'false-positive' results in a newborn screening program. *Pediatrics*, **73**: 183–187.
17. Bowling F, Cleghorn G, Chester A *et al.* (1988) Neonatal screening for cystic fibrosis. *Arch. Dis. Child.* **63**: 196–198.
18. Chatfield S, Owen G, Ryley HC, Williams J, Alfaham M, Goodchild MC, Weller P (1991) Neonatal screening for cystic fibrosis in Wales and the West Midlands: clinical assessment after five years screening. *Arch. Dis. Child.* **66**: 29–33.
19. Dankert-Roelse JE, te Meerman GJ, Martijn A, ten Kate LP, Knol K (1989) Survival and clinical outcome in patients with cystic fibrosis, with or without neonatal screening. *J. Pediat.* **114**: 362–367.
20. Weaver LT, Green MR, Nicholson K *et al.* (1994) Prognosis in cystic fibrosis treated with continuous flucloxacillin from the neonatal period. *Arch. Dis. Child.* **70**: 84–89.
21. Wilcken B, Chalmers G (1985) Reduced morbidity in patients with cystic fibrosis detected by neonatal screening. *Lancet* **ii**: 1319–1321.

22. Ranieri E, Ryall RG, Morris CP *et al.* (1991) Neonatal screening strategy for cystic fibrosis using immunoreactive trypsinogen and direct gene analysis. *Br. Med. J.* **302**: 1237–1240.
23. Dauphinais RM (1992) A cost analysis of blood-spot screening newborns for cystic fibrosis. *J. Clin. Immunoassay*, **15**: 121–125.
24. Boland C, Thompson NL (1990) Effects of newborn screening of cystic fibrosis on reported maternal behaviour. *Arch. Dis. Child.* **65**: 1240–1244.
25. Thelin T, McNeil TF, Aspegren-Jansson E, Sveger T (1985) Psychological consequences of neonatal screening for α_1-antitrypsin deficiency. *Acta. Paediatr. Scand.* **74**: 787–795.
26. McNeil TF, Sveger T, Thelin T (1988) Psychosocial effects of screening for somatic risk: the Swedish α_1 antitrypsin experience. *Thorax*, **43**: 505–507.
27. Zellweger H, Antonik A (1975) Newborn screening for Duchenne muscular dystrophy. *Pediatrics*, **55**: 30–34.
28. Firth M (1983) Diagnosis of Duchenne muscular dystrophy: experiences of parents of sufferers. *Br. Med. J.* **286**: 700–701.
29. Firth M, Gardner-Medwin D, Hosking G, Wilkinson E (1983) Interviews with parents of boys suffering from Duchenne muscular dystrophy. *Dev. Med. Child. Neurol.* **25**: 466–471.
30. O'Brien T, Sibert JR, Harper PS (1983) Implications of diagnostic delay in Duchenne muscular dystrophy. *Br. Med. J.* **287**: 1106–1107.
31. Gardner-Medwin D, Bundey S, Green S (1978) Early diagnosis of Duchenne muscular dystrophy. *Lancet*, **i**: 1102.
32. Gardner-Medwin D (1983) Recognising and preventing Duchenne muscular dystrophy (editorial). *Br. Med. J.* **287**: 1083–1084.
33. Smith RA, Rogers M, Bradley DM, Sibert JR, Harper PS (1989) Screening for Duchenne muscular dystrophy. *Arch. Dis. Child* **64**: 1017–1021.
34. Fenton-May J, Bradley DM, Sibert JR, Smith R, Parsons EP, Harper PS, Clarke A (1994) Screening for Duchenne muscular dystrophy. *Arch. Dis. Child* **70**: 551–552.
35. Smith RA, Williams DK, Sibert JR, Harper PS (1990) Attitudes of mothers to neonatal screening for Duchenne muscular dystrophy. *Br. Med. J.* **300**: 1112.
36. Bradley DM, Parsons EP, Clarke A (1993) Experience with screening newborns for Duchenne muscular dystrophy in Wales. *Br. Med. J.* **306**: 357–360.
37. Parsons E, Bradley D (1994) Ethical issues in newborn screening for Duchenne muscular dystrophy: the question of informed consent. In: *Genetic Counselling: Practice and Principles* (ed. A. Clarke), pp. 95–112. Routledge, London and New York.
38. Parsons E, Bradley D, Clarke A (1996) Disclosure of Duchenne muscular dystrophy after newborn screening. *Arch. Dis. Child* **74**: 550–553.

Chapter

9

PRENATAL GENETIC SCREENING

Paradigms and perspectives

A.J. Clarke

INTRODUCTION

As a preliminary to this discussion about prenatal genetic screening, I must re-emphasize the crucial distinction between population screening tests and individual diagnostic tests. The difference between a population genetic screening programme on the one hand, and diagnostic genetic testing offered in the context of the individual or family on the other hand, has already been set out in the chapter on carrier screening programmes. In the prenatal context, the contrast is between a screening test that is offered to every pregnant woman or couple simply because of the pregnancy, and a diagnostic test that is offered to a specific woman or couple because it is already known that a child born to them is more likely than usual to be affected by a specific genetic or congenital disorder; for example, there may be a family history of a genetic disease.

A screening test may in itself also be a diagnostic test, as when an ultrasound scan identifies a fetus with spina bifida. Other screening tests – such as maternal serum screening for Down syndrome, or simply putting the question "How old will you be at the expected date of your baby's birth?" to a pregnant woman – are probability-modifying tests rather than diagnostic tests. This is a second type of contrast to be drawn between the words 'screening' and 'diagnostic' which must also not be overlooked. A third contrast is that screening tests are actively offered by health professionals to large numbers of people who have not requested testing, while family-based diagnostic tests are provided to individuals who have actively sought information and testing for themselves.

In this chapter I am proposing to examine the issues raised by the population-based programmes of 'routine' prenatal genetic screening. Such tests are often regarded as routine because so many women participate that it may be unusual for a woman not to do so. Therein lies one of the principal problems of such programmes, the fact that they are often routinized in practice so that neither the staff offering the test nor the woman complying with the programme need stop and reflect "On balance, is this test going to be helpful?" Given the constraints of time available, it will often be impossible to initiate a lengthy discussion with every woman in antenatal clinic about every possible screening test, but obstetricians and midwives providing antenatal care may also feel that they will be criticized, and perhaps be exposed to litigation, if they do not make available every such test to every pregnant woman.

In the relative luxury of genetic counselling clinics, providing a service to small numbers of individuals and families who are referred because of some pre-existing concern about a condition suspected as running in the family, the time allotted to each consultation is much more generous than in an antenatal clinic; and detailed questioning and discussion can be encouraged. It is therefore all too easy for clinical geneticists like myself to criticize practice in obstetrics and midwifery, because we do not have to cope with such large numbers of clients. In so far as there is criticism of others in this chapter, it is meant to be constructive.

Many obstetricians and midwives are well aware of the difficulties that surround prenatal screening programmes, and participate in research to help improve practice. They are looking for solutions to the real problems they face, through identifying and promoting key aspects of good practice. Furthermore, many of the problems have arisen because the screening programmes have their origin in an area outside the usual scope of obstetrics and midwifery. The impetus to develop and promote prenatal screening has often come from 'public health genetics', which is now a small specialty but which has historic roots in the eugenics movement. While public health genetics is a thoroughly legitimate discipline, and while a number of its practitioners take care to distance themselves from the disreputable connotations that linger on from its past, some programmes have been established in which the driving force has been the desire to save money through reducing the birth incidence of certain 'costly' disorders. The danger of subordinating clinical practice to population goals, especially in relation to reproduction, is very real; it has happened in the past in a number of European countries and in North America, and is currently practised in China. Such abuses are discussed elsewhere in this book, but we must retain an awareness that we are separated from such practices ourselves by only a very fine line [1–3].

GOALS AND CONSEQUENCES

An important distinction must be drawn between the consequences of a genetic screening programme and its goals. In the area of prenatal genetic screening, there is the difference between recognizing terminations of pregnancy as a consequence of screening on the one hand, and counting terminations of pregnancy as the goal of screening (the principal measure of successful outcomes) on the other. To understand the ethos of a screening programme, it is important to define how it views both the selective termination of pregnancies affected by the condition for which screening has been made available and the birth of children affected by the same condition.

It is possible to examine the (often complex) relationship between the stated goals of a genetic screening programme, its ethos as manifest in its operating practices, and the measures of outcome that have been adopted to justify its continued existence or funding. The adoption of the 'prevention' of cases of a disorder (by means of selective terminations of pregnancy) as a principal goal may not be stated explicitly, but it may function as such if the ethos of the programme in practice leads towards the routinization of the test and the assumption that an affected pregnancy will be terminated. Terminations of pregnancy may also function as the principal, albeit implicit, goal of a programme if its evaluation or audit is centred upon counting the terminations of affected pregnancies that result from the programme. This situation may arise, for example, if the programme evaluation assesses the impact on the birth incidence of the condition [4].

We know from empirical research that staff in antenatal clinics frequently offer prenatal screening in a directive manner, clearly expecting their women clients to comply with the routine procedures of the clinic. Tacit programme goals may be transmitted by the health professionals [5–8]. Furthermore, the information provided about the diseases for which screening is offered, may be biased by the fact that the leaflets or other sources of information are being used in the context of prenatal screening [9]. We know from this evidence as well as from personal and clinical experience, from talking with friends, patients and colleagues, and from reading the written accounts of women who have regretted their participation in prenatal screening programmes, that such practices are widespread.

It is also important to examine the various public representations of prenatal screening programmes – the 'need' for prenatal screening as constructed both in the antenatal clinic and in society at large [10]. The public face of prenatal screening has important implications for women who become caught up in the consequences of such screening programmes, but there are wider implications as well which can be considered in three categories. There are the consequences of prenatal screening for all pregnant women, and so for all potentially pregnant women; widespread prenatal screening may have altered the whole experience of pregnancy, the relationship of every woman with her fetus [8]. There are consequences of screening for those individuals affected by the condition for which screening is offered – those individuals who would never have been born if they had not 'slipped through the net'. This is likely to have a real impact upon the self-esteem of already vulnerable individuals and may also increase the stigmatization of affected individuals and their families [11]. A third set of implications of prenatal screening are the social effects that follow from the inequitable access of different sections of society to a high quality of antenatal care. If prenatal screening is utilized more by women from higher income groups, then children with physical and mental disabilities may be born more frequently in poorer social classes and among ethnic minority groups, and these groups have less access to good social, medical and educational support for those with special needs.

In weighing up the arguments for and against the adoption of prenatal screening programmes, or in deciding what type of screening to adopt, all these consequences of screening must be examined in addition to the formal outcome measures.

GOALS OF PRENATAL SCREENING

There are three established candidates as to the principal goal for programmes of prenatal screening: sparing resources through the termination of pregnancies affected by disorders for which care is costly, the avoidance of suffering in affected children through the termination of affected pregnancies, and the promotion of informed reproductive decisions. None of these goals, I would maintain, is adequate to justify screening programmes, and each of them raises further problems which must be weighed against the putative benefits of screening.

If the justification of prenatal screening is to spare public resources – or to minimize health care costs in a private insurance system – then we must ask why prenatal screening, almost alone of all health care activities, is expected to make a profit. It is accepted that most health care activities consume resources, but that this is worthwhile because of the benefits that result. Simply saving money is not an adequate justification for an activity that has so many and such profound consequences. The rationale underlying such cost–benefit analyses – that society should be spared the burden of caring for individuals unworthy of life – amounts to much the same as the values underlying the very worst excesses of the eugenics of the 1930s.

If pushed to the absurd, a policy of prenatal screening run on this principle should aim to terminate any pregnancy where the resulting child would be likely to cost society more (in health care, housing, education, social benefits, etc.) than he or she would be likely to contribute in taxes over their lifetime. And what implications of this policy will there be for the ethos of prenatal screening? It would clearly be necessary to present compliance with prenatal screening as a duty with which society expects every woman to comply, and there would have to be sanctions against those who failed to do so. These sanctions could perhaps be financial; society might refuse to pay for the care of children with costly disorders whose births could have been 'prevented'. And what implications would this have for our treatment of the elderly and of prisoners? Such a policy is not ethically sustainable.

The avoidance of suffering is clearly a worthy goal in most circumstances. Why is it not suitable as a goal for prenatal screening programmes? First, there is a danger that this goal will effectively collapse into the first one, with similar consequences for the ethos of the programme – which will routinize screening, leading women to become caught up in it before they are given the chance to consider the issues fully. There is a second problem, that the suffering to be prevented may not so much be of the future child, as of the parents and family. The meaning of the word 'suffering' is then rather different from the concept of physical pain and distress experienced by the child, and becomes the emotional distress of the parents at having a child with a condition causing disability. While children born with spina bifida often do suffer in the physical sense, those with anencephaly probably do not suffer much in any sense and those with Down syndrome often do not suffer physically. Because the promotion of prenatal screening for Down syndrome may actually promote the stigmatization and intolerance that is a major cause of the suffering experienced by many affected individuals and their families, it is not at all clear that such screening is helpful.

The distress of parents whose child is born with any genetic disease or disability is very real, and I am not dismissing this factor as unimportant. The avoidance of this distress, however, is not commonly presented as the rationale for prenatal screening. One possible reason for this is the great difficulty of weighing the distress of having an affected child against the distress caused by prenatal screening, leading to the termination of a wanted pregnancy, especially when the problems resulting from prenatal screening affect many more individuals than the few who would have affected children in the absence of screening. In the different context of prenatal diagnosis performed because of a family history of a specific disease, the probability of a child being affected is much greater, the initial anxiety about the child being affected is present before the offer of any medical intervention, and the family will usually have practical experience of the disorder in question upon which to base their decisions. Although these decisions will always be difficult, they will be less rushed and will be based on firmer knowledge and experience than in the context of screening. Indeed, a parental decision to choose prenatal diagnosis and to terminate an affected pregnancy when they already know so much about the condition, can be understood as the parents vicariously taking upon themselves the distress of these procedures so as to spare their child the suffering of the disease.

The third potential goal of screening, the promotion of informed reproductive decisions, is also problematic when the concept is analysed. While any decisions that are to be made in this sphere of reproduction clearly should be made by individuals who are informed of the facts and have been offered help to confront the relevant issues, the simple maximization of such decisions is not a coherent goal [11–13]. This is a principle that must be respected, but it is not an objective in its own right. In a screening programme, reproductive decisions are being imposed upon a population; they can become a burden rather than a blessing. The imposition of this burden has clearly altered the whole experience of pregnancy in Western society, giving rise to the experience of the 'tentative pregnancy': the feeling that a pregnancy is only provisional until the fetal quality control investigations have been reported as satisfactory [8]. Until then, at 18–20 weeks' gestation, a woman may find it difficult to let herself feel unconditionally positive about her pregnancy. What implications does this have for the ability of parents, and also society as a whole, to welcome and care for any child, and especially for those children who do have serious problems despite all investigations? There will always be such special children who require even more loving care than the rest.

This goal of promoting informed reproductive choice has been proposed by Modell and Kuliev as a way of countering the objection made to some reproductive genetic screening programmes that they count success in terms of the number of terminations of pregnancy achieved, and that this is unacceptable [14]. Where a couple choose to face the risk of an affected child, and where this decision has been made on an informed and considered basis, then that would be counted as a success in this model. This is clearly a major advance when compared to crude eugenic notions of prenatal screening, but it is still inadequate.

If the maximization of informed reproductive decisions is itself a benefit, then it will be difficult to justify restricting the scope of such decisions. How serious does a disease or disability have to be to justify screening and possibly the termination of a pregnancy? In Britain, society has decided that prenatal sex selection and the termination of pregnancies on the grounds that the parents prefer a child of the other sex is not acceptable. Because most sex selection has resulted in the termination of female fetuses, our society has rejected this as being offensive to women. So how much offence must prenatal screening cause to people with Down syndrome before it, too, becomes unacceptable? Or what about screening for conditions that are largely cosmetic, such as cleft lip, or for non-disease characteristics? Promoting such screening would increase the numbers of informed reproductive decisions being made, but that does not make these tests worthwhile. The severity of the condition for which screening is proposed, and the effect of screening upon affected individuals and their families, must be taken into account when society makes decisions in this area.

A further reason for regarding the model of Modell and Kuliev as unsatisfactory is that it could, in less scrupulous hands than theirs, amount merely to a change in the packaging of prenatal screening without any real change in practice or ethos. It would be all too easy for a cynical eugenicist to adopt the verbal formulations of Modell and Kuliev as a shelter behind which to carry on the real business of 'genetic cleansing'.

At this point, we should examine the word 'prevention' in the context of prenatal screening. The 'prevention' of any disease is obviously desirable, but in this context the only conditions that can now really be prevented are neural tube defects (preventable by folic acid supplements in the mother's diet). What so often passes as prevention – the termination of a pregnancy carrying an affected fetus – is not prevention at all. I would argue that the word 'prevention' should only be used to refer to true, primary prevention and not to describe programmes of selective terminations of affected pregnancies. The euphemistic misappropriation of the word 'prevention' results in confusion and makes it more difficult to discuss these topics rationally.

PARADIGMS OF PRENATAL SCREENING

'Informed choice' vs. 'genetic cleansing'

We now turn to consider a way of evaluating programmes of prenatal screening, by attempting to locate a programme upon a spectrum between two possible extreme types (paradigms) of prenatal screening programme. These contrasted paradigms may be described as the 'informed choice' model and the 'genetic cleansing' model. It is likely that most screening programmes in Britain contain some features of both, and very few will be located at either extreme.

It is possible to utilize three types of information about a programme in order to evaluate it in this way, giving us three different perspectives on each programme: its stated goals, its implicit goals as it operates in everyday practice, and its implicit goals as revealed in its choice of outcome measures used to describe and justify itself. This approach to the evaluation of prenatal screening was developed during discussions some years ago that I had with my colleagues Drs Helen Hughes and Evelyn Parsons (Cardiff) and Dr Jo-Anne Finnegan (Toronto), and I am very grateful to them for their permission to incorporate some of these ideas into this chapter. By comparing a programme's evaluation from the three different perspectives, one can assess the consistency of the values it has adopted or espoused: does its practice match its rhetoric?

We can compare a number of current and proposed programmes with these two 'ideal types'. It is possible to construct a hypothetical continuum based on the underlying values of each programme. This allows comparisons to be made between programmes, and also allows the three aspects of each programme to be compared. At one end of the continuum are programmes aimed at preventing the birth of affected individuals in the most cost-effective manner; at the other end are programmes aimed at maximizing client autonomy by facilitating informed choice.

The motivation for this exploration is the concern that the formal and informal adoption of 'prevention' as a goal in genetic screening has led to clinical practices that conflict with the widely espoused value of respect for autonomous client decision-making. An editorial in the *Lancet* [15] called for organizers of carrier screening programmes for cystic fibrosis to "make explicit the objectives of the programme" because there is a considerable difference between establishing a screening programme to allow informed choices about

Table 9.1. Paradigms of prenatal screening

Issues	Spectrum: 'Cleansers'	Spectrum: 'Choosers'
Goal	Reduction in birth incidence	Provision of information and choice
Economic considerations	'Cost-effectiveness'	'Costs permitting'
Consent	Assent or opt-out	Fully informed
Pre-test counselling	Minimal	Unlimited and integral
Post-test counselling and support	Minimal	Unlimited and integral
Client satisfaction	Irrelevant	Studied systematically
Social evaluation	Nil or cursory	Careful search for problems
Focus of audit evaluation of programme	Selective abortion	Informed, autonomous, reproductive choice

reproduction, and establishing a programme to prevent the birth of affected babies and thereby save money for the taxpayer and/or improve the 'genetic fitness' of the population. This applies more widely to other genetic screening programmes, and such analysis may be able to focus attention on areas where deficiencies in existing programmes can be identified, and to influence the development of future screening programmes so that clinical practice will adhere more closely to our common values (*Table 9.1*).

THE TWO 'IDEAL TYPES'

At one end of the continuum of possible screening programmes are those which aim primarily to reduce the birth incidence of the conditions for which screening is offered. Whether the espoused goal of the programme is to save money or to minimize suffering, any such programme will be driven by cost considerations (financial audit), to achieve the highest rate of termination of affected fetuses (euphemistically referred to as 'secondary prevention') for the least expenditure. At this 'cleanser' extreme, there would be a willingness to withhold information from those screened or offered screening, if this was thought to be in the best interests of the programme. The provision of pre- and post-test counselling and support would be a low priority and might be neglected because of cost, because education and support are not programme goals or because pre-test counselling might lower the rate of compliance with the programme.

The concern to achieve a high uptake rate for the test could lead to its portrayal as an established routine, with only occasional clients 'opting out' of the test; those who opt out would be portrayed as irresponsible, and would be subject to clear professional disapprobation. The same concern could also lead to the test being portrayed as a means of providing 'reassurance' to anxious, expectant parents. The fact that the screening programme cannot possibly provide complete reassurance about the well-being of the future infant may not be made explicit, and the fact that the anxiety which the test is said to allay may well have been provoked by professionals seeking to persuade women to accept the screening test may also be ignored [16]. Furthermore, women will not be asked in advance how they would feel if an unfavourable or uncertain result was obtained from screening; they will not be encouraged to think that the screening test might provide anything other than reassurance.

In contrast, the alternative paradigm places emphasis on providing information, support and reproductive choice, and values the autonomy of clients. This type of 'chooser' programme provides pre- and post-test information and counselling, and may involve health education in schools and the wider community, although this would not be tightly linked to the offer of testing. Genetic testing is only performed on the basis of informed consent, and women are not encouraged to participate so as to gain 'reassurance' [16,17]. All clients will be provided with complete results, unless they have specifically requested otherwise. The programme is designed so that anxiety and regrets generated by testing are kept to a minimum, by encouraging reflection and discussion in advance and by providing information about a test at an early stage – long before a decision need be made. Ongoing support would be available for all who are offered screening. Programme evaluation is an integral part of its operation, and monitors how the test is offered, the social composition of those recruited, their pre-test preparation, the reproductive decisions they make, their social and psychological responses and their satisfaction with the service as a whole.

Of course no real prenatal screening programme will fit either caricature. It is not possible here to offer a detailed evaluation of even a single programme, but published reports of prenatal screening from the past decade will be cited to illustrate how such an evaluation can operate and to demonstrate that this method of analysis can be helpful.

PRENATAL SCREENING FOR DOWN SYNDROME

A cost–benefit analysis of prenatal screening for Down syndrome was published in 1976, when the method analysed was to offer amniocentesis to pregnant women over a certain age. It was shown that offering amniocentesis in this way to women over 40 years would produce a net saving of public resources, and the costs of offering amniocentesis to women over 35 years would be balanced by the savings [18]. Yet this paper did not recommend the adoption of such a programme of prenatal screening for Down syndrome. It concluded with this passage:

> The findings also re-emphasise that society's response to the problem of Down's syndrome cannot rest solely on considerations of economic costs and benefits. If

> Down's syndrome is socially unacceptable, provision of a programme to reduce its birth prevalence by scarcely a third would be an inadequate response. Conversely, failure to implement a programme for all maternal age groups would imply that there were other, perhaps more appropriate, responses to the problem of Down's syndrome. Since this would call into question any programme directed at identification and termination of affected pregnancies, it would be logical to resolve this dilemma before any programme was started.

Yet this has not happened – society has baulked at this discussion, and has never debated these issues openly. While there may be a broad view that the termination of pregnancies affected by serious genetic diseases should be legally permitted, there is no consensus that terminating affected pregnancies is the most appropriate response to the challenge of Down syndrome and that this view should be promoted publicly through the National Health Service. This debate has scarcely even started, despite some bold attempts [19–21], except within the confines of religious groups or single-interest pressure groups. Instead, prenatal screening for Down syndrome and other conditions has crept in quietly through the work of a coalition of professional screening enthusiasts drawn from obstetrics, genetics and public health. The caution of Hagard and Carter has not been heeded; perhaps they should have given it more prominence.

Most publications from the last 10 years that deal with prenatal screening for Down syndrome – now employing biochemical measures of markers in maternal serum – focus on the proportion of Down syndrome fetuses that could be identified by a given method and then terminated, with success clearly being assessed by the potential reduction in birth incidence of infants with Down syndrome or by the financial savings to society or to the health services (or both) [22–27]. They lean heavily towards the 'cleanser' paradigm of screening. The only outcomes given serious consideration in these papers – more than lipservice – are the tangible, economic outcomes. This is the most cynical form of genetic reductionism: the reduction of life to pounds or dollars.

Over the same period, the idea that such screening is a necessary part of health services has been promoted in the media by enthusiasts who have substantial vested interests in the widespread adoption of screening. In California, legislation has led to the forceful promotion of serum screening – women who do not wish to participate are required to provide a written statement of refusal. The portrayal of screening to pregnant women as a way of providing reassurance and of reducing risk, when in fact it does neither, has become accepted by institutional interests that approve of screening upon other grounds; this has led to the almost universal adoption of screening without an open public debate on the real issues [28,29].

While serum screening has been introduced on a very wide scale, it has become clear that there are grave problems associated with it, which are scarcely mentioned in the glowing reports written from the perspective of the 'cleansers'. There is the finding that a positive serum screening test leads on to significant maternal anxiety, and that this is largely resolved by a normal result from amniocentesis, but that anxiety persists throughout the pregnancy and beyond it in those who do not choose to face the emotional trauma and risk of miscarriage associated with amniocentesis [30]; once on the conveyor belt of screening, a result indicating increased risk forces a choice between putting the pregnancy at risk or persistent anxiety. There is the evidence that many obstetric services are offering serum screening with inadequate knowledge and resources to counsel their clients properly, and this leads to confusion and distress in many women [31–34] – especially in those given a result indicating an increased probability of Down syndrome, or in whom an unanticipated diagnosis has been made (a result that is neither Down syndrome nor completely 'normal') [35]. One factor that may exacerbate the problems, is that many of those providing counselling in relation to screening for Down syndrome, and even many genetic counsellors, may have insufficient knowledge or experience of Down syndrome [36].

Other techniques of screening for Down syndrome can be considered, such as the combination of serum screening with fetal echocardiography or the use of early ultrasound measurement of fetal nuchal thickness, but these approaches do not resolve any of the issues of principle: the value judgements that are usually implicit but which require public airing. Further research is required to document some of the other negative consequences of promoting widespread screening for Down syndrome, including the unwillingness of parents to care for a child born with Down syndrome despite normal screening tests, and other effects on affected individuals and their families [37].

FETAL ULTRASOUND ANOMALY SCREENING

In a 'chooser' fetal ultrasound programme, every effort will be taken to ensure that the client understands the purpose of the test and its limitations, what types of results are possible (including false-positive and false-negative results), and what decisions she might then be asked to make; the possibility of the screening process leading to a termination of pregnancy would be indicated at the outset. The woman will be encouraged to think through what her response might be to a scan result suggesting fetal abnormality before agreeing to have the scan, and she may be asked to provide written consent for the procedure as in the case of amniocentesis or chorionic villus sampling. The promotion of the scan as a 'routine' procedure, undertaken 'for reassurance' or 'to promote bonding', will not be encouraged in this type of programme but may be promoted under the 'cleanser' approach; for example, the sale of ultrasound images as 'Your First Picture of Baby' may be promoted because it functions as a means of maximizing the uptake of ultrasonography.

To characterize any particular ultrasound anomaly scanning service, a set of questions could be asked: what provision is made for pre-scan information and counselling? What type of consent is sought before scanning? What training do the ultrasonographers (medical or otherwise) have in communication skills? Is the woman encouraged to attend the scan with her partner? When an anomaly is found, is it suggested that the couple meet with a paediatrician or surgeon to discuss possible treatment? What measure of outcome is chosen for audit: correct diagnoses made (with confirmation at pathology or at delivery)? Or terminations 'achieved'? What steps are taken to inform the primary health care team when an anomaly is identified, and what follow-up is provided? (i.e. is proper support provided for women and their partners who undergo termination of pregnancy for such fetal abnormalities? and, is this costed as part of the screening?).

There are many problems with ultrasound scanning for the detection of fetal anomalies as practised in Britain. First, like serum screening for Down syndrome, it has been introduced in a piecemeal fashion by enthusiasts, and without any coherent evaluation policy [38]. Also, it has been introduced without the incorporation of proper counselling into the service, so that the general standard of pre-scan counselling has been described as extremely poor [39]. There is also some evidence that staff involved in providing fetal anomaly scans actively resist the provision of information to their clients, particularly if this leads to some women declining the offer of screening [40]. Finally, it must be remembered that fetal ultrasound scans may be understood by parents and professionals in quite different ways. To parents, they may be an opportunity to enter whole-heartedly into parenting. To professionals they are a form of surveillance or quality control, with the option of terminating the pregnancy always just around the corner [41]. It is no wonder that it can be difficult for staff to explain *their* purpose in offering scans, and nor is it surprising if the parents become distraught when these two perspectives collide – when a serious fetal anomaly is discovered.

One reason for adopting a policy of fetal anomaly scanning would be if it led to improvements in perinatal mortality. While such improvements have not been found in some series [42,43], they have been reported in the Helsinki ultrasound trial [44,45]. This improvement, however, may not represent any real gains because it is largely accounted for by terminations of pregnancy in which the fetus had a lethal malformation; if that group is excluded then the effect is no longer apparent. The improvement in outcomes is effectively cosmetic, with the shift of inevitable deaths from the perinatal period to before 28 weeks' gestation. This may be doing the woman no favours, however, because it imposes a burden of guilt and remorse on the mother and her partner, and may make it more difficult for them to receive social support from relatives, friends and professionals. The need for support following such terminations of pregnancy [46–48] and the long-lasting distress, depression and social isolation for both the woman and her partner [49] have now been well documented. Saari-Kemppainen *et al.* [44,45] provide no discussion whatsoever of the counselling issues surrounding fetal ultrasonography and the possibility of adverse outcomes after 'successful' diagnoses.

Another study of the efficacy of routine prenatal ultrasonography for the detection of fetal structural abnormalities [50], despite having the technical efficacy of the procedure as its focus, does refer to the counselling provided by an obstetrician and a midwife counsellor whenever an abnormality is identified. It is also clear that follow-up support is offered to these women. It is acknowledged that high levels of anxiety can be caused by fetal ultrasonography, and may be prolonged if the interpretation of the ultrasound findings is uncertain. It is suggested that the way to avoid such problems, however, is to screen more populations and collect more data – a purely technical solution to the problem. There is also a clear expression of the intention to reduce the perinatal mortality by increasing the number of fetuses terminated because of lethal malformation: " . . . the potential for reducing perinatal mortality may be even greater, as in six cases the parents elected not to intervene despite the identification of a potentially lethal fetal abnormality." This inference that lower perinatal mortality figures automatically amount to positive health gain is unjustified, as discussed above. Despite this spurious assumption, however, it can be seen that this screening programme presents itself as being further from the 'cleanser' paradigm than do the reports from Helsinki.

A third study of the effectiveness of routine ultrasound scanning " . . . in terms of accuracy of detection of fetal structural abnormality and the effect on obstetric and neonatal care" to be considered here is that by Luck *et al.* [51]. While reducing perinatal mortality by terminating malformed fetuses in the second trimester is regarded uncritically as a health gain, mention is made of the need for 'adequate counselling facilities', and emphasis is placed on the prompt re-examination by a consultant radiologist of women in whom a suspicious scan result has been found. Even more important, and unlike the other two papers, reference is made to pre-scan counselling by each woman's consultant obstetrician which was reiterated by the radiographer before any scan was performed. This study therefore is notable for its emphasis on pre-scan discussion and informed consent. Indeed, about 4% of women in the project declined the offer of scanning. Although the programme has a goal of terminating fetuses with malformation, it does recognize an important role for education and consent.

The comparison of these published reports cannot permit us to draw any conclusions about how the programmes operate in practice, but do give us the chance to compare how they portray themselves. It would be interesting to compare the programmes in operation, to see whether this paper comparison matches reality. At least this exercise may have clarified the use of the paradigms of prenatal screening to characterize screening programmes. It must not be forgotten that the antenatal diagnosis of some fetal anomalies may lead to improved outcomes for the affected infants, especially in the case of certain gastrointestinal and cardiac malformations, and so there may be outcomes of anomaly scanning other than termination of an affected fetus.

PRENATAL CARRIER SCREENING PROGRAMMES

A final topic to consider is the question of prenatal screening for genetic carrier status. Any carrier screening programme is oriented towards reproductive decisions, but the offer of carrier screening in pregnancy makes this particularly obvious. Given that pregnant women find it difficult to decline the offer of prenatal diagnostic tests [11,52], the decision to offer carrier screening for recessive diseases in the antenatal clinic must convey to staff and clients a powerful, albeit perhaps implicit, message. The impression must be given that 'society', and in particular the medical profession, are promoting and recommending the screening programme with a view to selective terminations of pregnancy.

It would not be appropriate to discuss carrier screening here at great length, because many of the issues are dealt with in other chapters (those on carrier screening and on outcomes), but it is perhaps useful to look at the issues surrounding couple testing in antenatal clinics. The fact that testing has been offered in pregnancy at all could undermine the confidence of the screened couple that ongoing support would be available to care for an affected child if they declined the offer to terminate an affected pregnancy. This would be especially true in an insurance-based health care system, but may even apply in countries with a national health service (and this just about still includes Britain). That such concerns are not merely my imagination is clear from the careful documentation of genetic discrimination in North America [53–55]. The very offer of this type of carrier screening in pregnancy indicates that the programme is closer to the 'cleanser' end of the continuum than the 'chooser' end. The high acceptance rate of prenatal carrier screening when it is offered early in pregnancy by the woman's general practitioner could be accounted for by the women feeling that the test is of special relevance to them at that time or by their finding it especially difficult to decline testing when it is offered in that context [56]. It is not clear which explanation is nearer the truth.

One way in which prenatal carrier screening for cystic fibrosis (CF) has been offered is to couples [57], with the result being given to the couple as either 'screen-positive' (both individuals identified as carriers) or as 'screen-negative' (at least one individual not identified as a carrier). This means that couples in which only one individual is identified as a carrier are not given that information, and the chance of the couple having an affected child is small but is greater than the probability they are given – the pooled probability of all those apart from the couples in which both individuals are shown to be carriers. Conversely, the chance of an affected child being born to couples in which neither individual is shown to be a carrier will be even smaller than the probability they are given.

This approach obviously offends the sensibilities of those – like myself – who consider that, in general, tests should not be performed if the results will be withheld from the individuals concerned. It also denies members of a carrier's extended family the opportunity to seek genetic counselling, and it may provide false reassurance that neither member of the couple is a carrier; this could be especially important if the carrier member of the couple later found a different partner. One reason for adopting this policy, however, may be that many fewer couples will be worried by the information that one of them is a carrier than would be the case if all identifiable carriers were given their results during a pregnancy. For CF, this could amount to one in 10–15 couples, if both partners were tested in all couples. Which approach, then, is better – couple testing with the withholding of information, or testing of carrier status with full disclosure of results which will alarm many women and couples? The 'cost-effectiveness' of different approaches to screening will vary between populations [58], but the judgement as to which to adopt is not simply a matter of economics [59]. Of course, one could conclude from this that neither approach is justifiable.

If carrier screening is worthwhile for reasons other than the termination of affected pregnancies then testing with the full disclosure of results would be preferred; if the primary goal of screening is to terminate affected pregnancies then couple screening might be regarded as superior. The goals of couple screening as set out by Wald were defined in the opening statement of the 1991 paper:

> The main aim of screening for congenital disorders is the prevention of handicap with the least harm from side-effects and at the lowest cost for a given level of prevention [57].

This statement is admirable for the clarity with which it locates the goals of the programme as the cost-effective prevention of genetic disease. Potential psychosocial problems are briefly referred to in terms of anxiety and worry, but the conclusion is drawn that antenatal couple screening is 'practical' and the 'most effective' method, since the antenatal clinic is a ' . . . natural "screening turnstile"'. One problem with such a perspective is the tone of the information provided about the disease in question – CF. This may be of great importance, because the portrayal of the disease 'to be prevented' must influence the decision of individuals as to whether or not they would wish to undergo testing. Is the clinical information provided to those unacquainted with the disorder influenced by considerations of the economic cost of treatment? This issue arises because the greater the success achieved in CF clinics in prolonging and in improving the quality of the lives of CF sufferers, the greater the financial 'benefits' of terminating CF pregnancies. On the other hand, and paradoxically, the better the prospects for CF children born now, the less will be the advantage of 'prevention' as perceived by the families. Unsurprisingly, Wald took a moderately pessimistic view of the current life expectancy of newborn individuals with CF [60,61]. If an effective, rational, gene-based but expensive therapy becomes available, as it may, so that the outlook for affected individuals improves still further, then tension will increase in the prenatal context between clinical optimism and considerations of cost to the health service.

Finally, mention must be made of proposals to introduce population-based prenatal screening to identify female carriers of the fragile X syndrome [62]. This raises numerous social and ethical problems, even assuming that the very real technical problems of distinguishing between carrier females and non-carriers can be overcome; there actually may not be a sharp distinction. Is it really better to identify carriers during a pregnancy, when decisions must be rushed and pressured, as opposed to identifying affected individuals and then offering genetic counselling to their families? Which course should we choose if we cannot provide both services? What effect will such a population-based approach have on individuals with fragile X syndrome – some of whom are scarcely affected at all? What problems are likely in a prenatal screening programme given that we cannot predict the severity of an affected individual, especially a female carrier? Some fragile X female carriers have a mild degree of learning problems, and it may be especially difficult to ensure that the quality of the informed consent to participate (or not) is of a high quality for these individuals; this group may be especially vulnerable to being led compliantly into a screening programme which they later regret unless very great care is taken.

CONCLUSIONS

Prenatal screening programmes for genetic conditions and congenital malformations do not meet the standard criteria for screening programmes [63], because the usual intervention made available when an affected fetus is identified is a termination of pregnancy. This is clearly not 'therapy' in the usual sense of the word, and is not in any sense therapeutic for the affected fetus. While prenatal diagnosis and the termination of affected pregnancies may well be considered appropriate when it is made available within a specific family context in which the probability of an affected fetus is great and the family is actively seeking prenatal tests, the social context of prenatal population screening is very different. It cannot simply be assumed that screening offered to a population at low risk of a condition, and who are not actively seeking testing, will also be experienced as helpful.

Professional enthusiasm for a new technology is not sufficient to justify its adoption – as has happened with the introduction of CF carrier screening in Denmark [64] and elsewhere. Such enthusiasm has combined with 'public health' considerations to create a strong coalition of interests that have dominated the organization of antenatal health care and that have largely controlled the portrayal of prenatal screening in the media. The possible harmful consequences of prenatal screening have received relatively little attention, and such evaluations of prenatal screening as have been carried out have neglected some of the most problematic issues. Of particular importance is the ethos of a screening programme, which may be very difficult to capture in a neat verbal formulation, but which is so important to the experiences of the women and families who come through the system.

The ethos of a clinical service may be part of what makes the difference between the experiences of women who are given genetic information by professionals from different backgrounds. We know that this makes a considerable difference to the decisions women make as to whether or not to terminate a pregnancy affected by a sex chromosome abnormality; women given the information by a geneticist are less likely to terminate the pregnancy than those who are given the information by an obstetrician [65,66]. More generally, it is known that the framing of health professionals' statements about reproductive risk, and not just the level of risk, has a considerable influence on the woman's perception of the risk [67]. Given this evidence, it is clear that the active promotion of prenatal screening, and the professionals' assumption that an affected pregnancy should be terminated, will have profound effects on the women in their care.

The use of narrow, financial considerations to justify prenatal screening is common [68–71], but is fortunately being challenged from within health economics: it can be pointed out that such cost–benefit analyses depend upon the implicit judgement that screening for the relevant disorder, and the termination of affected pregnancies, is socially desirable. Thus: "economics should not be used to decide whether or not to have a prenatal screening programme: this is a political decision" [72]. Without open discussion about what programmes *should* be available, it will not be possible for health professionals – let alone society as a whole – to work towards a consensus on the manner in which prenatal screening should be implemented and the extent to which it should be promoted.

The offer of reproductive choice itself, as argued above, is not sufficient to justify imposing the burden of complex and unwelcome decisions on so many people, nor is this in any sense a convincing goal when espoused by genetic epidemiologists or public health physicians. It evades the question of who is deciding what choices to put to the public, and it ignores the cost of providing such services. Furthermore, in operation, a screening programme that defined the birth of a wanted, affected child as a benefit rather than a burden might appear indistinguishable from one that regarded any affected child as a cost. If the practical operation of such a programme is largely untouched, the redefinition of successful outcomes could be regarded as being primarily of cosmetic significance.

We now need to build on the preliminary studies of prenatal screening programmes to understand the consequences of participation for everyone involved, whatever the results and whatever the outcomes of the pregnancy. With some exceptions, medical professionals (including geneticists) have made little effort to examine the social and ethical implications of prenatal screening programmes [73]. Given the way in which health professionals influence both public attitudes to genetic diseases and the reproductive decisions of individual women, it is disingenuous of clinicians to claim that they are simply making available services that the public manifestly wants to utilize. In fact, they are frequently generating the 'needs' that they then claim to meet, and they are choosing to find a 'solution' to this 'need' rather than to other needs that may be just as urgent and rather more amenable to simple, benign interventions [11,64,74].

We need to attend more seriously to the 'side-effects' of this construction of needs on pregnant women. We need to attend to the 'side-effects' on those individuals with the disorders that 'need to be prevented' and on those with other forms of disability – the self-esteem of those with disabilities and the willingness of society to tolerate and support them. We need to look actively for other evidence of possibly harmful consequences of prenatal screening that have so far received little attention because many such consequences will be missed by the usual methods of assessing screening outcomes – the 'hidden casualties' of screening. And we need to consider other ways in which reassurance can be provided to parents in relation to the possible adverse outcomes of a pregnancy [75]. Society then needs to reconsider what types of prenatal screening it is *really* helpful to make available.

ACKNOWLEDGEMENTS

I would like to thank many colleagues for their support and encouragement in the train of thought developed here. In particular, I would like to thank Drs Evelyn Parsons, Helen Hughes and Jane Fenton-May in Cardiff, Dr Jo-Ann Finnegan in Toronto, Dr Carina Wallgren-Pettersson in Helsinki and Prof. John Burn in Newcastle-upon-Tyne, England.

REFERENCES

1. Harper PS (1992) Genetics and public health. *Br. Med. J.* **304**: 721.
2. Duster T (1990) *Backdoor to Eugenics*. Routledge, New York.
3. Paul DB, Spencer HG (1995) The hidden science of eugenics. *Nature*, **374**: 302–304.
4. Clarke A (1990) Genetics, ethics and audit. *Lancet*, **335**: 1145–1147.
5. Marteau TM, Slack J, Kidd J, Shaw R (1992) Presenting a routine screening test in antenatal care: practice observed. *Publ. Hlth*, **106**: 131–141.
6. Marteau TM, Plenicar M, Kidd J (1992) Obstetricians presenting amniocentesis to pregnant women: practice observed. *J. Reprod. Infant Psychol.* **11**: 3–10.
7. Rapp R (1988) Chromosomes and communication: the discourse of genetic counselling. *Med. Anthr. Quart.* (New Series) **2**: 143–157.
8. Rothman BK (1988) *The Tentative Pregnancy: Prenatal Diagnosis and the Future of Motherhood*. Pandopra Press, London.

9. Lippman A, Wilfond BS (1992) Twice-told tales: stories about genetic disorders. *Am. J. Hum. Genet.* **51**: 936–937.
10. Lippman A (1991) Prenatal genetic testing and screening: constructing needs and reinforcing inequities. *Am. J. Law Med.* **17**: 15–50. Reprinted in *Genetic Counselling: Practice and Principles* (ed. A. Clarke), pp. 142–186. Routledge, London.
11. Clarke A (1991) Is non-directive genetic counselling possible? *Lancet*, **338**: 998–1001.
12. Chadwick R (1993) What counts as success in genetic counselling? *J. Med. Ethics*, **19**: 43–46.
13. Clarke A (1993) Response to: "What counts as success in genetic counselling?" *J. Med. Ethics*, **19**: 47–49.
14. Modell B, Kuliev AM (1991) Services for thalassaemia as a model for cost–benefit analysis of genetics services. *J. Inher. Metab. Dis.* **14**: 640–651.
15. Anonymous (editorial) (1992) Screening for cystic fibrosis. *Lancet*, **340**: 209–210.
16. Green J (1990) *Calming or Harming: a Critical Review of Psychological Effects of Fetal Diagnosis on Pregnant Women*. Galton Institute, London. Galton Institute Occasional Papers, Second Series, No. 2.
17. Marteau TM (1990) Screening in practice: reducing the psychological costs. *Br. Med. J.* **301**: 26–28.
18. Hagard S, Carter FA (1976) Preventing the birth of infants with Down's syndrome: a cost–benefit analysis. *Br. Med. J.* **i**: 753–756.
19. Boston S (1981) *Will, My Son. The Life and Death of a Mongol Child*. Pluto Press, London.
20. Boston S (1994) *Too Deep For Tears*. Pandora (Harper Collins), London.
21. Williams P (1995) Should we prevent Down's syndrome? *Br. J. Learn. Disab.* **23**: 46–50.
22. Wald NJ, Cuckle HS, Densem JW *et al.* (1988) Maternal serum screening for Down syndrome in early pregnancy. *Br. Med. J.* **297**: 883–887.
23. Sheldon TA, Simpson J (1991) Appraisal of a new scheme for prenatal screening for Down's syndrome. *Br. Med. J.* **302**: 1133–1136.
24. Wald NJ, Kennard A, Densem JW *et al.* (1992) Antenatal maternal serum screening for Down's syndrome: results of a demonstration project. *Br. Med. J.* **305**: 391–394.
25. Piggott M, Wilkinson P, Bennett J *et al.* (1994) Implementation of an antenatal serum screening programme for Down's syndrome in two districts (Brighton and Eastbourne). *J. Med. Screening*, **1**: 45–49.
26. Shackley P, McGuire A, Boyd PA *et al.* (1993) An economic appraisal of alternative pre-natal screening programmes for Down's syndrome. *J. Pub. Hlth Med.* **15**: 175–184.
27. McColl AJ, Gulliford MC (1993) Population health outcome indicators for the NHS: a feasibility study. *Population Needs and Genetic Services: an Outline Guide.* Faculty of Public Health Medicine, HMSO, London.
28. Browner CH, Press NA (1995) The normalization of prenatal diagnostic screening. In: *Conceiving the New World Order: the Global Politics of Reproduction* (eds F.D. Ginsburg, R. Rapp) pp. 307–322. University of California Press, Berkeley, CA.
29. Press NA, Browner CH (1994) Collective silences, collective fictions: how prenatal diagnostic testing became part of routine prenatal care. In: *Women and Prenatal*

Testing: Facing the Challenges of Genetic Technology (eds K.H. Rothenberg, E.J. Thomson) pp. 201–218. Ohio State University Press, Columbus, OH.
30. Marteau TM, Cook R, Kidd J (1992) The psychological effects of false-positive results in prenatal screening for fetal abnormality: a prospective study. *Prenat. Diag.* **12**: 205–214.
31. Statham H, Green J (1993) Serum screening for Down's syndrome: some women's experiences. *Br. Med. J.* **307**: 174–176.
32. Green JM (1994) Serum screening for Down's syndrome: experiences of obstetricians in England and Wales. *Br. Med. J.* **309**: 769–772.
33. Smith DK, Shaw RW, Marteau TM (1994) Informed consent to undergo serum screening for Down's syndrome: the gap between policy and practice. *Br. Med. J.* **309**: 776.
34. Dennis J, Sawtell M, Rutter S (1993) Counselling needed after screening for Down's syndrome. *Br. Med. J.* **307**: 1005.
35. Fry V, Rustad J (1993) Antenatal screening for genetic disease. *Br. Med. J.* **307**: 1498.
36. Baumiller RC (1992) A candid look at Down's syndrome (book review) *Am. J. Hum. Genet.* **51**: 938–939.
37. Julian-Reynier C, Aurran Y, Dumaret A *et al.* (1995) Attitudes towards Down's syndrome: follow up of a cohort of 280 cases. *J. Med. Genet.* **32**: 597–599.
38. Chitty LS (1995) Ultrasound screening for fetal abnormalities. *Prenat. Diag.* **15**: 1241–1257.
39. Smith DK, Marteau TM (1995) Detecting fetal abnormality: serum screening and fetal anomaly scans. *Br. J. Midwif.* **3**: 133–136.
40. Oliver S, Rajan L, Turner H *et al.* (1996) Informed choice for users of health services: views on ultrasonography leaflets of women in early pregnancy, midwives, and ultrasonographers. *Br. Med. J.* **313**: 1251–1255.
41. Sandelowski M (1994) Channel of desire: fetal ultrasonography in two use-contexts. *Qual. Hlth Res.* **4**: 262–280.
42. Bucher HC, Schmidt JG (1993) Does routine ultrasound scanning improve outcome in pregnancy? Meta-analysis of various outcome measures. *Br. Med. J.* **307**: 13–17.
43. Ewigman BG, Crane JP, Frigoletto FD *et al.* (1993) Effect of prenatal ultrasound screening on perinatal outcome. *N. Engl. J. Med.* **329**: 821–827.
44. Saari-Kemppainen A, Karjalainen O, Ylostalo P, Heinonen OP (1990) Ultrasound screening and perinatal mortality: controlled trial of systematic one-stage screening in pregnancy. *Lancet*, **336**: 387–391.
45. Saari-Kemppainen A, Karjalainen O, Ylostalo P, Heinonen OP (1994) Fetal anomalies in a controlled one-stage ultrasound screening trial. A report from the Helsinki Ultrasound Trial. *J. Perinat. Med.* **22**: 279–289.
46. Lloyd J, Laurence KM (1985) Sequelae and support after termination of pregnancy for fetal malformation. *Br. Med. J.* **290**: 907–909.
47. Kenyon SL, Hackett GA, Campbell S (1988) Termination of pregnancy following diagnosis of fetal malformation: the need for improved follow-up services. *Clin. Obstet. Gynecol.* **31**: 97–100.
48. Elder SH, Laurence KM (1991) The impact of supportive intervention after second trimester termination of pregnancy for fetal abnormality. *Prenat. Diag.* **11**: 47–54.

49. White-van Mourik MCA, Connor JM, Ferguson-Smith MA (1992) The psychosocial sequelae of a second-trimester termination of pregnancy for fetal abnormality. *Prenat. Diag.* **12**: 189–204.
50. Chitty LS, Hunt GH, Moore J, Lobb MO (1991) Effectiveness of routine ultrasonography in detecting fetal structural abnormalities in a low risk population. *Br. Med. J.* **303**: 1165–1169.
51. Luck CA (1992) Value of routine ultrasound scanning at 19 weeks: a four year study of 8849 deliveries. *Br. Med. J.* **304**: 1474–1478.
52. Sjogren B, Uddenberg N (1988) Decision-making during the prenatal diagnostic procedure. A questionnaire and interview study of 211 women participating in prenatal diagnosis. *Prenat. Diag.* **8**: 263–273.
53. Billings PR, Kohn MA, de Cuevas M *et al.* (1992) Discrimination as a consequence of genetic testing. *Am. J. Hum. Genet.* **50**: 476–482.
54. Holtzman NA, Rothstein MA (1992) Eugenics and genetic discrimination. *Am. J. Hum. Genet.* **50**: 457–459.
55. Geller LN, Alper JS, Billings PR *et al.* (1996) Individual, family and societal dimensions of genetic discrimination: a case study analysis. *Sci. Eng. Ethics*, **2**: 71–88.
56. Harris H, Scotcher D, Hartley N *et al.* (1996) Pilot study of the acceptability of cystic fibrosis carrier testing during routine antenatal consultations in general practice. *Br. J. Gen. Pract.* **46**: 225–227.
57. Wald NJ (1991) Couple screening for cystic fibrosis. *Lancet*, **338**: 1318–1319.
58. Cuckle HS, Richardson GA, Sheldon TA, Quirke P (1995) Cost effectiveness of antenatal screening for cystic fibrosis. *Br. Med. J.* **311**: 1460–1464.
59. Miedzybrodzka ZH, Hall MH, Mollison J *et al.* (1995) Antenatal screening for carriers of cystic fibrosis: randomised trial of stepwise vs couple screening. *Br. Med. J.* **310**: 353–357.
60. Britton J, Knox AJ (1991) Screening for cystic fibrosis. *Lancet*, **338**: 1524.
61. Elborn JS, Shale DJ, Britton JR (1991) Cystic fibrosis: current survival and population estimates to the year 2000. *Thorax*, **46**: 881–885.
62. Palomaki GE (1994) Population based prenatal screening for the fragile X syndrome. *J. Med. Screen.* **1**: 65–72.
63. Wilson JMG, Junger G (1968) *The Principles and Practice of Screening for Disease*. Public Health Papers 34. World Health Organisation, Geneva.
64. Koch L, Stemerding D (1994) The sociology of entrenchment: a cystic fibrosis test for everyone? *Soc. Sci. Med.* **39**: 1211–1220.
65. Holmes-Siedle M, Ryyanen M, Lindenbaum RH (1987) Parental decisions regarding termination of pregnancy following prenatal detection of sex chromosome abnormality. *Prenat. Diag.* **7**: 239–244.
66. Robinson A, Bender BG, Linden MG (1989) Decisions following the intrauterine diagnosis of sex chromosome aneuploidy. *Am. J. Med. Genet.* **34**: 552–554.
67. Shiloh S, Sagi M (1989) Effect of framing on the perception of genetic recurrence risks. *Am. J. Med. Genet.* **33**: 130–135.
68. Gill M, Murday V, Slack J (1987) An economic appraisal of screening for Down syndrome in pregnancy using maternal age and serum alpha-fetoprotein concentration. *Soc. Sci. Med.* **24**: 725–731.
69. Wald NJ, Cuckle HS (1987) Recent advances in screening for neural tube defects and Down syndrome. *Baillières Clin. Obstet. Gynaecol.* **1(3)**: 649–676.

70. Tosi LL, Detsky AS, Roye DP, Mordern ML (1987) When does mass screening for open neural tube defects in low-risk pregnancies result in cost savings? *Can. Med. Ass. J.* **136**: 255.
71. Goldstein H, Philip J (1990) A cost–benefit analysis of prenatal diagnosis by amniocentesis in Denmark. *Clin. Genet.* **37**: 241–263.
72. Phinn N (1990) *Can Economics be Applied to Prenatal Screening?* Discussion Paper 74, Centre for Health Economics, York.
73. Lippman A (1991) Research studies in applied human genetics: a quantitative analysis and critical review of recent literature. *Am. J. Med. Genet.* **41**: 105–111.
74. Farrant W (1985) Who's for amniocentesis? The politics of prenatal screening. In: *The Sexual Politics of Reproduction* (ed. H. Homans). Gower, Aldershot.
75. Lippman A (1992) Mother matters: a fresh look at prenatal genetic testing. *Issues Reprod. Genet. Eng.* **5**: 141–154.

Section III

GENERAL ISSUES IN GENETICS

The chapters in this section cover a variety of topics that do not fit readily under any single heading, but which all raise issues which we feel are both important and timely. They include topics involving both genetic services and genetic research.

Chapter 10 takes up the theme of the previous section, population screening in genetics, and looks at it from the perspective of public health professionals. It raises the tensions, both ethical and practical, between population-directed and individual-directed goals in medical genetics; if the main role of clinical geneticists is to relate to individual families, then who should be responsible for the wider population issues and how can we be sure that such people are aware of and do not forget the needs of individuals? The chapter begins directly with a short paper already published, which still seems to us to provide a clear introduction.

Chapter 11 examines the challenges to genetic privacy that arise in clinical genetic practice. Family members and third parties may all have reasons to be interested in learning about the genetic constitution of specific individuals. Where an individual wishes to protect their genetic privacy, conflict may ensue. How should we meet such challenges to medical confidentiality?

Chapter 12 develops the related theme: how should we define our goals and our measures of outcome. We need to do this accurately and objectively if we are to justify health-care funding for genetic services, yet all who have approached this topic have found it an extremely difficult one. As can be seen in the chapter, it is possible to define which outcome measures are inappropriate or harmful, but much harder to decide what might be appropriately used in their place.

In the next chapter, Chapter 13, the process of genetic counselling is explored. The emphasis that has been put upon non-directiveness as a goal of genetic counselling can be explained – it serves several valid functions – but we need to develop a richer description of the process of genetic counselling than simply to score it as directive or non-directive.

The final two chapters in the section, while written quite independently by the two of us, have proved to be closely linked and are the only two in the book which deal with genetic research rather than service aspects. As a consequence we are primarily addressing a different community – those involved in basic research who are not involved with service provision, nor perhaps having any clinical contacts with patients and families. We are aware that the view expressed in both chapters, that there are certain sensitive areas of research which should only be approached with the greatest caution (or possibly not at all) is one that will be opposed on principle by some scientists. Certainly the history of science would suggest that attempts to ban particular fields of work are not likely to be productive or even feasible. Yet we would hold that the other extreme, of such sensitive areas of research as behaviour genetics being pursued without consideration of their social consequences or of likely public reaction, is equally unacceptable, and that it is the responsibility of the scientific community to take the lead in promoting a responsible and socially acceptable approach to basic research in this and other equally sensitive fields.

Chapter 10

GENETICS AND PUBLIC HEALTH

P.S. Harper

The first part of this chapter appeared as a 'personal view' in the *British Medical Journal*[a]; it is followed by a commentary that updates the situation and links it to the area of 'genetic screening' already discussed in some detail in the previous sections of this book.

OVERVIEW

Genetic disorders and their prevention have become one of the principal challenges in medicine. As a clinical geneticist, I am excited by the major advances in isolating major human disease genes and am proud that we are able to apply these rapidly and often effectively to help families with serious genetic disorders. I feel uneasy, however, when I look at the future possibilities for misuse of these advances. In the past few years, a field known as 'community genetics' has emerged. This represents the population-based rather than individual family-orientated aspects of medical genetics, notably screening programmes for genetic disorders and for congenital malformations in pregnancy. The proliferation of such programmes demands co-ordination and initial evaluation and I have been among those who have supported this. So why am I ambivalent and worried?

Preclinical testing for genetic disorders is not new, but the development of an increasing number of accurate and totally age independent tests for genetic disorders is going to pose major problems, which have already been encountered for disorders such as Huntington's disease. At present, such tests are mainly undertaken in the context of family risk, but the increasing possibility of testing for specific mutations is shifting their use towards a diagnostic and population basis. Should we test all apparently sporadic cases of a disorder such as Alzheimer's disease to detect the minority that could be caused by inherited mutations? The response of the public health physician might

[a]Reproduced from the *British Medical Journal* (1992), Vol. 304, p. 721, by kind permission of the BMJ Publishing Group.

depend on the yield of such a test (currently uncertain), but of greater importance might be the effect on the family of identifying a serious risk to the mental health of relatives in later life when this was previously unsuspected. If screening for such mutations becomes a part of regular practice it will have major implications, comparable with those of screening for HIV.

The only molecular test to have actually entered the population screening arena so far is that of carrier screening for cystic fibrosis (see Chapters 5 and 6), the common mutation for which is carried by around one in 25 individuals in populations of northern European origin. Here some of the problems of carrier screening are lessened by the recessive nature of the disease – carriers are and will remain clinically normal – but will the perception of those detected reflect this?

Experience from haemoglobinopathy and Tay–Sachs carrier screening suggests that avoidance of stigmatization and other problems depends on the effectiveness of education regarding the disorder and its carrier status. Is our population sufficiently educated to be able to handle this information fruitfully? It will be important for this to be resolved by the current pilot projects before any general programme is accepted as a service.

Occupational testing is an area where molecular genetics may in future impinge directly on public health medicine. As susceptibility genes are progressively identified for common cancers, cardiovascular disorders and mental illness, should they be tested for in the context of environmental hazards or stress of particular occupations? If so, how will such information be used by employers and others? An allied and more immediate problem relates to the use of genetic testing in life and health insurance, already discussed in Chapter 4. Again the experience of HIV testing in this context provides an important and none too encouraging lesson.

Testing children for late-onset genetic disorders is a subject that has worried my colleagues and me. I can think of no more important issue which has been so ignored by those professionals who ought to be most concerned. At present, this problem arises principally in the context of individual family risk (Chapter 2), but the potential for population applications already exists. Every baby has a filter paper blood spot taken for detecting such treatable disorders as phenylketonuria and hypothyroidism, but what rules should govern its use for detecting late-onset and untreatable disorders? The potential for harm resulting from the ill considered testing of children for adult genetic disorders is great, yet most clinical colleagues are reluctant to admit it as a major issue.

How can we best prevent future problems? We need to distinguish between population and individual aims. Population prevention of a serious genetic disorder (e.g. Huntington's disease), is a valid goal to aim for, but it is different from the aim of genetic counselling and any associated predictive testing, where the goal is to help a specific individual, couple, or family achieve what is the optimal decision for themselves. To avoid such possible conflicts of interest it could be argued that those undertaking the individual genetic counselling of families should not be those also running population-based programmes. I would recommend the reverse, since each will gain insight by being exposed to the potentially opposing point of view. Thus if any defined field or subspecialty of community genetics does develop, I think that those concerned will benefit by having some commitment to seeing and counselling individual families rather than simply regarding them as numbers.

To avoid problems in the future, we need to be aware of the past. Most of those working in medical genetics (including myself until recently) are ignorant of the past abuse of our subject. It is easy to regard the excesses of the eugenics movement, or the abuses in Nazi Germany (Chapters 12–14), as disconnected from present day medical genetics, but a closer look at these episodes shows that their key feature was the subordination of individual decisions to the broader population-based goals. That these goals were often deeply flawed is not the relevant point. Those concerned at the time believed that they were acting in the best interests of the present and future population and that these must override the lesser rights and decisions of individuals. There is a clear danger here for how we judge the success of any population-based genetic programmes, and for conflict with any individual decisions that seem to jeopardize such success.

Finally, future developments should be discussed and debated openly, involving the public as well as professionals. A major problem in carrying this out in Britain, as in most countries, is the poor state of education of the public regarding science in general and genetics in particular. The tendency of the media to highlight the sensational and threatening aspects does not help. It is therefore important for those engaged in community genetics to undertake such education in relation to screening programmes and the broad area of genetic disorders before new programmes are introduced.

The field of medical genetics is now at a stage where it has the potential for influencing whole populations, not just the families at high risk. The potential for harm is correspondingly great if we do not think carefully about the effects of new developments. We should be able to avoid most of the serious problems if we do not forget the individuals who make up all populations. Any applications of genetics in public health need to retain the respect for the autonomy of individuals and for the inevitable variation in decisions that people make. It will be a challenge during the coming years for medical geneticists and those in public health to combine their respective skills so that individual families and entire populations benefit from the discoveries in genetics that are likely to continue and even increase in pace over the next decade.

POSTSCRIPT

It is now 5 years since this short paper was written and published as a 'personal view' in the *British Medical Journal*. The main issue involved was, and is, the inherent tension between the needs and benefit of the individual and that of society in general; this tension is reflected in the differing goals of doctors involved with individual patients and those responsible for the health of populations – public health medicine.

There is no doubt that almost all medical geneticists see their primary duty as being to the individual family, but most would also feel a sense of broader responsibility to the general population, in particular to ensure that the services they provide are available to all at need, not just to the minority who have the education, and in many countries the finance, to obtain them.

This natural concern, common to all fields of medicine but particularly appropriate to medical genetics where the family, often extended, is the principal focus, leads to a direct interface with the field of public health medicine, and in particular to programmes of screening for genetic disorders, which will inevitably detect individuals or individual families who need detailed genetic counselling and management.

Who should actually be responsible for these genetic screening programmes and for ensuring that they are appropriately linked to the more specific and detailed genetic services already established for individuals? In 1992 the term 'community genetics' was in use for this general area of activity, and the suggestion made in my paper was that a specific group of professionals might emerge who would be visibly distinct from clinical geneticists on the one hand, and from public health physicians on the other. I implied that such professionals would need to be led by a clinician who had training in both these areas, and also suggested that some continuing 'hands on' commitment in clinical genetics might help to prevent such specialists forgetting that all populations are ultimately made up of individuals.

With hindsight, this suggestion has proved to be premature, probably to be naive and possibly impractical, since essentially nothing has happened in this direction during the intervening time and no subspecialty of 'community genetics' has appeared nor looks like doing so, either in Britain or elsewhere. Indeed, some professional screening experts have insisted that there is no need for genetic screening to be considered different in any way from other screening and that geneticists need not be involved. Thus, the editor of the new *Journal of Medical Screening*, in an editorial [1] entitled 'what is genetic screening anyway?', stated,

> Another reason for avoiding the term 'genetic screening' is that it may imply that *geneticists* should carry out screening that falls under this heading, as screening relies on 'genetic' tests. The terminology should be neutral with respect to who does the screening; often screening for genetic disorders requires little specialist genetic expertise.

Although agreeing with much of what this editorial said, I felt that it needed some response and made the following rejoinder [2][a].

The term 'genetic screening' is indeed often used in a confused and inaccurate way. The word 'genetic' should be equally applicable to screening based on a gene product as well as on DNA technology, and it may well be that we see the pendulum swing back in this direction in future. It is also often not made clear whether the object to be screened is a gene (e.g. screening for mutations in the cystic fibrosis gene) or a population (e.g. screening pregnant women for cystic fibrosis carrier status). Frequently 'screening' is inappropriately used for testing of individuals or families, not broader groups, where the issues may be very different.

I also agree with the view expressed that the fact that a screening programme is genetic in nature should not necessarily imply that geneticists should undertake or co-ordinate it; neither clinical nor laboratory geneticists are usually trained in the relevant epidemiological approaches, while the primary role of clinical geneticists is to help individual patients and family members. In any case, the number of geneticists is too small for this to be feasible generally.

I disagree however, with the implication that screening for genetic disorders carries no particular implications or issues that differ from other forms of screening. It does, and the failure of many epidemiologists to recognize this, related in part to absence of genetics in their training, can result in significant problems that an awareness of genetics and willingness to work closely with geneticists should help to avoid. The foremost difference is that the genetic nature of a disorder often results in risk implications to family members of the person screened, even though they may not be, nor perhaps wish to be, included in the screening programme. This may highlight the difference in goals of a 'public health' orientated screening programme and the more individual or family-based approach of genetic counselling and testing services.

Comparable problems may arise in a research context; thus epidemiological studies of Creutzfeldt–Jakob disease some years ago raised potentially serious issues when genetic testing for prion mutations was applied to a research population without initial recognition of the profound consequences for identifying such a mutation in a specific family. Now that mutations are detectable in subsets of many diseases not previously recognized as having an important genetic contribution (e.g. breast and colon cancer), the setting up of screening programmes without genetic input could well cause far more serious problems than if such programmes were initiated by geneticists without involvement of epidemiology and public health.

The answer seems clear – screening is of necessity a multidisciplinary process, and where genetic diseases are concerned it is wise to involve closely those who have clinical and laboratory genetic expertise. More generally, in light of the likelihood that genetic technology will indeed increase the feasibility of and pressure for 'genetic screening', not all of it justified, it seems important

---

[a]Reproduced courtesy of the *Journal of Medical Screening*.

and urgent that those involved in epidemiology and public health medicine should have a substantial component of genetics in their training, rather than the token amount now included in the UK. This would allow them to give specific and informed criticism, where this is needed, to a greater extent than they are able to at present.

This final point in my *Journal of Medical Screening* response on the need for education in genetics for those in epidemiology and public health medicine is in my view the key issue, for in Britain there is virtually no genetics included in the training of these specialties, whether medical or otherwise. In my own institution, with well established public health medicine training courses, I was recently asked to 'cover genetics' in a 45 minute talk! We thus have the situation that those involved with determining public health medicine programmes, and who also have a wider influence on commissioning of health care, including genetic services, are almost completely ignorant of even the simplest elements of genetics, and have little concept of what medical genetics as a specialty is or does. It is perhaps not surprising that fundamental points of relevance to families with genetic disorders may never be considered by those initiating and supervising 'genetic' screening programmes.

Where does this leave us in 1997? Not much further forward in my view, and with the need for an effective interface between the 'individual' family practice of clinical genetics and the population view of epidemiology and public health medicine as great as ever. I still feel that the preferable solution is for every health region – the level at which many population programmes are co-ordinated – to have one of its public health medicine staff interested and experienced in medical genetics, who could then have special responsibility for the growing number of population-based programmes. But how will this happen if such workers have no training in genetics? – especially if their seniors are reluctant to admit that it is a subject of importance.

Despite this, I am not entirely pessimistic about the situation; medical genetics has proved to be a remarkably collaborative and flexible specialty – who could have predicted effective joint programmes with surgeons in the area of familial cancer? I look forward to seeing a group of people emerge, under whatever name, who can help to ensure that the benefits of medical genetics services really do reach the full range of people who need them, and who can also help to plan screening and other population services for genetic disorders in a way that does not remove free choice or cause harm in any way to individual families.

REFERENCES

1. Wald NJ (1996) What is genetic screening anyway? *J. Med. Screen.* **3**: 57.
2. Harper PS (1996) What is genetic screening? *J. Med. Screen.* **3**: 165–166.

Chapter

11

CHALLENGES TO GENETIC PRIVACY

The control of personal genetic information

A.J. Clarke

INTRODUCTION

Genetic information about one individual may be important to others. Genes are shared within a family, of course, so information about one person's genetic constitution may be of real relevance to other members of the family for several reasons (concern about the individual and their partner or children, and perhaps concern about themselves). Insurance companies, employers and other third parties such as the state may also have an interest in the genetic constitution of individuals. Insurance companies and employers may wish to know, for example, how likely we are to develop a serious illness before the age of 60 or 65 years. Biotechnology corporations might be interested to know of any disease predisposition we have that could be modified by drugs, and the state might want to confirm our identity in relation to crime or paternity disputes.

Given that personal genetic information about one individual may be of interest to others, then the issue arises as to how private this personal genetic information should be. When individuals wish to disclose information about themselves to other family members or to third parties, then there is no problem. When they wish their genetic privacy to be respected, however, there may be a conflict with other parties who would like access to the information [1,2].

Respect for confidentiality has been one of the core principles of medical ethics from the time of Hippocrates. As one component of respect for autonomy, it is still central to the concerns of contemporary medical ethics. How strong, then, is the claim to respect for confidentiality in the context of genetic information? The General Medical Council sets the standards of professional conduct in Britain, and it permits the disclosure of personal information without consent in order to prevent serious harm to others: when would such circumstances apply in the context of clinical genetics? When do the claims of third parties or other family members override the individual's right to confidentiality?

To respect confidentiality in relation to genetics, the individual's consent should be obtained before any genetic information is disclosed to another family member or to a third party. This would include information about their clinical diagnosis, their genetic test results and their family structure. To be thoroughly consistent, it can be argued that respect for privacy also entails that consent should be obtained before genetic information about one individual is utilized in providing genetic counselling for other family members; we will return to this issue later.

THIRD PARTIES

Insurance companies will clearly be interested in any information about the risk of premature death or ill-health in their clients. Such information may in fact undermine the social basis of insurance, however, whose purpose is to spread across society as a whole the financial burden that is imposed on families by the unanticipated death or ill-health of young or middle-aged adults. The ability to individualize risks of ill-health may weaken this collective response to insecurity, and could perhaps even weaken respect for institutions such as the National Health Service (NHS). These issues have been discussed elsewhere [3,4] and are examined further in Chapter 4.

Employers may also be interested in the likely future health of their employees; both their general well-being and their susceptibility to specific occupational diseases. Discrimination against job applicants who are somewhat more likely than average to develop a serious illness in middle life could lead to the development of a genetic underclass of healthy individuals who would be denied employment or confined to low-paid jobs because of a disease susceptibility that might never manifest. There is already evidence from the USA that discrimination against certain individuals or families by insurance companies restricts their employment opportunities [5–7].

There may be more justification for screening for susceptibility to specific occupational diseases, and tailoring the work of a susceptible individual to minimize their exposure to the relevant environmental agent. It would not be acceptable, however, for this approach to justify a relaxation of safety standards for the workplace or the environment as a whole on the basis that the susceptible individuals were no longer exposed to the hazard. While some individuals may be more susceptible to damage from exposure to workplace hazards, no employee should be exposed to hazards that could reasonably be reduced or eliminated [8]. Society as a whole has an interest in workplace safety and in minimizing the release of toxic substances into the environment.

The disease, α_1-antitrypsin deficiency (α_1AT), predisposes adults to chronic lung disease, especially if they smoke or if they are exposed to dust in their workplace. This is still the only occupational disease for which genetic screening is currently feasible. An analysis of the practical and ethical issues raised by genetic screening for α_1AT concluded that the identification of symptomatic workers and attention to the control of the environmental triggers of the disease are more appropriate than a genetic screening programme [9].

The intimate connection between employment and health insurance in the USA means that any difficulty in obtaining employment will make it more difficult to obtain health insurance, and any genetic test that indicated a risk of ill-health would jeopardize employment because an employer's insurance scheme may be unwilling to provide cover. Suitable policies to tackle these issues of discrimination have been developed in the USA [10], which would restrict the right of the employer to obtain or use genetic information about his employees. Other possible means of ensuring genetic privacy for citizens have been raised for discussion, including the provision of anonymous genetic counselling for certain diseases [11] (although this may merely distract attention from proper efforts to ensure that genetic privacy is consistently protected [12]) and the use of certificates of confidentiality [13], a legal device specific to the USA. The protection or development of a national health care system providing universal coverage is another approach that would ameliorate the effects of such discrimination; this requires the political will to substitute the rationing of health care on the basis of 'need' for the rationing of health care on the basis of personal wealth.

The state itself may have an interest in personal genetic information, especially in confirming identity for forensic purposes or for demonstrating biological relatedness in paternity or immigration disputes. The potential problems arising from state-held genetic databanks in the USA have been discussed at length [14,15]. The establishment of genetic databanks by the police forces in Britain has generated remarkably little protest or discomfort, perhaps because the scope of application of such databanks is portrayed as being limited to facilitating police investigations into cases of murder or rape. The mixed record of British governments' respect for the civil liberties of their citizens gives me no reassurance that other, less acceptable uses for such genetic information will not be found.

CONFIDENTIALITY WITHIN THE FAMILY

There are three related issues that arise in the context of genetic privacy within a family. First, there is the question of transmitting information about an individual's risk of developing or transmitting a genetic disorder in the future: do individuals have the right to withhold relevant information about themselves from other family members? Secondly, there is the question of an individual's right not to undergo genetic investigations when this might be helpful to other members of the family. Third, there is the question of the genetic testing of children. Testing children, when it will not be relevant to them until they are older, invades their privacy (by removing their right to confidentiality) and abrogates their autonomy (by removing their right to make their own decisions in the future). This is discussed in Chapter 2 and will not be considered further here.

Transmitting genetic information within the family

When an individual is found to be a healthy, unaffected carrier of a genetic disorder, they will be told that they may have children affected by the disorder in question. Indeed, they may only realize that they are carriers of a specific disorder as the result of the birth of an affected child. If the condition is autosomal recessive in inheritance, then a child can be affected only if both parents are carriers; if the condition is a sex-linked disease or is a balanced chromosomal rearrangement (usually a translocation), then a child will be at risk whatever the genetic constitution of the individual's partner (see Chapter 6 on carrier screening).

Once one member of a family has been identified as a carrier, then other family members may also be carriers in the same way. If they are informed about this possibility, then they will be able to seek genetic counselling to discuss their particular situation and also, if they so wish, to be tested to determine their carrier status. Once informed of the risk, individuals can avoid having any children or they can decide to avoid the birth of an affected child by prenatal diagnosis and the selective termination of affected pregnancies.

While many individuals and couples immediately sense the potential importance of this type of genetic information to other family members, others react differently: they may have an impulse to keep such information secret. Such impulses to be secretive are usually very short-lived, often being associated with an immediate urge to deny the unwelcome information they have been given. These impulses are usually soon replaced by the desire to ensure that other members of the family are fully informed of their own potential risk of transmitting the condition to their future children. When the desire for secretiveness persists, so that important information is not passed to other family members, then most genetic counsellors will ask the individual(s) concerned to think through the potential implications of this failure to pass on the information. This leads to consideration of a scenario in which another member of the family has a child affected by the disorder in question, realizes that it is genetic and works out that (s)he could have been warned about the possibility in advance. The possibility of bitter family disputes arising or being exacerbated is clear, and most individuals will not want to put family relationships at such jeopardy.

It is unusual for individuals to refuse to transmit important genetic information to other relevant family members, once they have had a chance to think through all the implications of such a decision. Careful follow-up of families with balanced translocations shows that most possible carriers can be informed and tested, especially if direct contact is maintained with the family by experienced professionals over some years [16,17]. Where family members are not willing to assist in this process, however, should genetic counsellors have a duty to override confidentiality and to pass on such information to the relevant family members?

It has been argued that health professionals do have this obligation [18]. The case for involuntary disclosure has been put forward in a report from the Royal College of Physicians of London [19], and the Nuffield Council on Bioethics has suggested that criteria to define the circumstances in which confidentiality may be breached should be considered and drawn up [8]. I wish to disagree with the principle of involuntary disclosure on a number of grounds [20], as others have done [21,22].

(i) To establish criteria for the circumstances in which medical confidentiality can, indeed must, be breached is a very serious matter. Currently, such breaches are permitted, under exceptional circumstances, to prevent a risk of death or serious harm to others. The very act of establishing criteria for breaching confidentiality, however, alters the event from being an exception, which must therefore be justified on every occasion, to itself being the rule. This will weaken medical respect for confidentiality in general, and this will be to the detriment of the whole medical profession.

(ii) The practical effects of diminishing medical respect for genetic privacy will have numerous consequences for the practice of clinical genetics – even of clinical medicine – because of its effects on the doctor–patient, or counsellor–counsellee, relationship. To ask for permission from an individual for their consent to pass on genetic information to others will be seen to be bogus, because it will be realized that the clinician would pass on the information in any case, with or without consent. Individuals with a family history of a serious genetic disease will become increasingly reluctant to reveal this even to their family doctors, and certainly to clinical geneticists, in case information about their private griefs are passed on to other family members without consent. Many more individuals will feel a strong, initial urge to privacy than would have declined to disclose important information to their relatives after careful consideration and appropriate counselling. This will lead to much more harm resulting from the undermining of respect for confidentiality than by the current policy of fully respecting confidentiality while encouraging openness within families.

(iii) Without the co-operation of the individuals who are refusing to pass on the information, it will often be quite impossible to trace and notify the other relevant members of the family. They may, indeed, live in another country or another continent. Efforts to trace them will often be doomed to failure, but attempts to trace such family members could effectively become an obligation, imposing a new and onerous duty of care on the whole profession of genetic counselling.

(iv) The reason for informing other members of the family of their situation is so that they can take informed and considered reproductive decisions. To impose a duty on families and professionals to pass on genetic information to others, rather than merely recommending it as good practice, would be to lend support to the concept of genetic harm. To justify a forced disclosure of genetic information entails arguing that *not* to pass on the information amounts to committing harm. The only type of harm that could seriously be invoked here would be the birth of a child with, or at risk of, a genetic disorder. This amounts to support for the concepts of wrongful birth or wrongful life, and courts in most jurisdictions have, with good reason, been very reluctant to give these concepts any legal force. To undermine professional respect for confidentiality and thereby to strengthen the validity of wrongful life or wrongful birth suits would be a very unfortunate development.

Similar considerations apply to passing on genetic information about late-onset genetic disorders in the family. The circumstances in which such concerns become relevant usually apply to the transmission of information from parents to their (adult) children about a condition that one of the parents has developed or may develop in the future. There are occasions where the at-risk or affected parent has only recently found out about the condition in the family – perhaps Huntington's disease or autosomal dominant polycystic kidney disease – and is still coming to terms with it. It is certainly all too possible for the problems that arise from a late, traumatic disclosure of such a diagnosis to be repeated in succeeding generations unless the affected or at-risk parent is able to summon up the moral courage to disclose the true situation to their children at an appropriate age [23]. If it was known that a geneticist could force a disclosure at a difficult time for the family, it would make many families reluctant to discuss the issues at all. Sensitive genetic counselling can, in fact, be very helpful in bringing families to share information and emotions more fully and frankly; heavy-handed interventions in such a process would be very damaging.

Where there are therapeutic or preventive interventions that are known to improve the medical outcomes for those at risk of a genetic disorder, the arguments in favour of an early disclosure to other family members are stronger. These considerations certainly apply to some familial cancer syndromes and to some forms of familial blood lipid disorders associated with premature coronary artery disease. These considerations are usually understood very clearly by all concerned in these families, and there are few cases of reluctance to disclose the relevant information. In these circumstances as well, it is clear that enforcing disclosure would lead to far more problems than arise with the present system of respecting confidentiality while encouraging disclosure and openness. The legal situation in this area of disclosure to other family members is uncertain, although the law as it is now probably favours the preservation of confidentiality within the doctor–patient relationship [24,25].

Consenting to genetic tests on behalf of other family members

There are circumstances in which it may be helpful to one family member for a relative to provide a blood sample for genetic studies. Such requests used to be made frequently when linkage-based genetic analyses were more usual, and samples often had to be obtained from a number of family members for there to be any possibility of genetic testing providing useful information. Essentially, a number of affected and unaffected relatives would be compared to see which copy of the relevant gene was associated with the disease in that family; this could be tracked through the family to see who else might be at risk of developing the condition in the future. Now, it would be more likely that a single blood sample might be requested from one affected member of the family to permit identification of their disease-causing mutation; once the mutation was identified, then predictive or prenatal testing would be available to any interested family member.

But what happens if one or more individuals, from whom a sample would help in the analysis, decide against providing the sample? It may be claimed that they have a moral obligation to provide the sample – in a particular set of circumstances – but should they also have an enforceable, legal obligation to do so?

This question has been raised for serious consideration in the report from the Royal College of Physicians [19], but needs to be countered. Not only may there be good arguments, at least arguments worthy of some respect, that support the individual refusing to provide a sample (they may not approve of the termination of pregnancy that might result from them providing the sample, or may consider the testing to be inappropriate at that time) but there may be other reasons underlying the refusal. For example, refusal might reflect the person's belief that family relationships are not as they seem (non-paternity may be an issue) or the individual may fear, not unreasonably, that a covert predictive test would be carried out on him/herself.

In addition to these possible grounds for refusal to provide a sample are the harms likely to be committed in the process of enforcing collection of a sample. Who would hold down a rational, healthy man or woman while a blood sample was taken against his or her wishes? Would it be the police? Or a private security firm? Or clinical geneticists? What doctor would be willing to commit such an assault? Would not the General Medical Council then have to consider the professional future of anyone involved in such a process, which appears to be an assault on fundamental human rights [26] as well as merely on the body? Another approach would be to obtain a sample by deception – on some plausible pretext – but that is also surely wholly unethical except in circumstances where the individual is incompetent to make a decision, where due legal process has determined that a sample may be obtained, and where an innocent deception (a 'white lie') would be the kindest way of achieving this.

In all of these circumstances, is it not clear that the genetic information about an individual, or residing in their blood cells, belongs to that individual in a stronger sense than that in which it belongs to other members of the family, with whom it may (or may not) be shared? To attempt to enforce the contrary view would lead to unpleasant absurdities that would be morally repugnant. Furthermore, the dependence of most genetic analyses on samples from other family members is likely to be temporary, soon to be outmoded by technical developments. Surely, it must be better to continue to place our trust in the guidance we have received from the Greeks.

Respect for autonomy?

It may appear strange that 'autonomy', in the sense of the control of personal genetic information, is being challenged at the same time as exaggerated respect is being paid to a different 'autonomy', in relation to individual reproductive decisions about possible future children. In fact, there is no contradiction. The meaning of 'autonomy' in relation to reproductive decisions is changing from the original notion of a free reproductive choice made on personal and moral grounds to a notion of free choice laden with responsibility for the legal and financial consequences of the birth of a child with a 'preventable' genetic disease. Instead of individual self-determination, an appeal to 'autonomy' has come to represent the imposition of self-reliance, as the shift of meaning of 'autonomy' in this context has been characterized by Chadwick [27]. Indeed, respect for autonomy is being replaced by a fear of the financial or legal penalties that may be meted out to those individuals or families who decline to follow the 'responsible', socially approved course of action. Confidence in society's willingness to support families caring for those with disability or disease has been undermined, and has led to this shift in meaning of 'autonomy'.

RESPECT FOR CONFIDENTIALITY IN GENETIC COUNSELLING PRACTICE

In the everyday practice of clinical genetics, there are opportunities for confidentiality to be breached inadvertently as well as deliberately. What can we practitioners do to help preserve confidentiality and respect for confidentiality in the course of genetic counselling? We certainly need to obtain the consent of our clients before:

(i) passing on information about them to other family members or to clinical genetics colleagues involved with other members of the family, or
(ii) utilizing information from their genetic records or test results in providing genetic counselling to other members of the family.

We also need:

(i) to take great care to avoid the *inadvertent* disclosure of information about one family member to another. Such inadvertent disclosures may occur if a family tree is constructed from information provided by one member or branch of a family and is then shown to other members of the family. Secrets may be disclosed, such as terminations of pregnancy, cases of false paternity or unacknowledged adoptions. The avoidance of such problems may entail the drawing and use of different versions of the family tree for each family unit within the overall extended family. Problems can arise even from the seemingly innocent custom of writing the names of family members referred for genetic counselling on the front of the family's genetics file; this can result in the disclosure to other family members that a specific individual has been referred to the genetics department;

(ii) to ensure that genetic tests *only generate the information that has been requested by the family*. It is easy for a laboratory to generate more information than is required to answer the diagnostic or predictive question that has been asked, and this will sometimes lead to problems that could have been avoided. This is particularly true in linkage-based investigations, and so is becoming less of a problem now that diagnostic testing relies more heavily on the detection of mutations. In family linkage studies, where the question of paternity may be critical to the interpretation of a test result, this should be discussed in advance with the relevant family members. *The most important way of avoiding the inappropriate disclosure of personal genetic information is not to generate unnecessary information in the first place*;

(iii) to ensure the clean separation between diagnostic genetic tests and research investigations. In the former, families can expect a concrete result within a defined time-frame; with the latter, there may be no result of any relevance to the family, and any result that does emerge may take months or years to do so and may be of uncertain significance. A research sample obtained years before for a genetic linkage study of a late-onset disorder could all too easily be used to carry out predictive testing as part of the research. A sample taken for research purposes should not be used in that way unless the original consent obtained clearly recognized such predictive testing, or unless fresh consent has been obtained from the individual concerned.

If such a result is generated in the course of research, however, it should certainly not find its way into the individual's medical records. That is why the separation between research and diagnostic samples should be very clear and should extend to the system of laboratory record-keeping as well as the clinical files. Without this separation, results will be generated and revealed inappropriately. These issues have already been discussed more fully by Harper [28] and others [29,30].

THE PROBLEMS OF REGISTERS

Another aspect of clinical genetic practice that can lead to particular unease is the existence of genetic registers. To keep a list of individuals with a genetic disorder, or at risk of developing it, certainly requires justification. If comprehensive, computer-based lists of this sort had existed in Germany during the 1930s, then they would surely have led to even worse abuses of patients than actually occurred [31] (and see other chapters in Section IV of this book). The existence of genetic registers today could be portrayed or understood as a system for the imposition of genetic counselling on those who do not want it: a system of tracking and control. This, I will seek to demonstrate, would be a misunderstanding, but the term 'genetic register' may have these connotations for some people. Another concern about genetic registers relates to the potential loss of genetic privacy; individuals in families with certain genetic diseases may fear the loss of confidentiality about their diagnosis or their situation of genetic risk, and may have concerns about access to that information by third parties. If sensitive information is stored on a hospital computer, will it be accessible to commercial 'hackers', for example, who could sell the information to employers or insurance companies?

Concerns about confidentiality and privacy are always to be taken seriously, and the security of hospital computer systems is certainly a cause for concern. What needs to be addressed here, however, is the meaning of the term 'genetic register'. I hope that this will allay many inappropriate concerns, and allow us all to concentrate on the areas where debate is necessary and may lead to improvements in practice.

What is a genetic register?

The term, 'genetic register', in fact has no single, clear meaning. It covers a multitude of different entities that have evolved independently and for a great range of different purposes. I will list the different types known to me.

1. *A list of patients referred to a genetics service.* This is really an appointments system, although it may contain some clinical details such as the presumed diagnosis of the condition in the family. If unauthorized access was obtained to the computer record through the NHS network, it could indeed be very damaging to have a diagnosis of Huntington's disease or familial cancer listed against someone applying for insurance or employment, especially when the diagnostic label might well not apply to them but to a relative by blood or even by marriage.

Such a register raises no specific problems simply because it relates to genetic diseases, and the issue of confidentiality needs to be resolved for non-genetic conditions as well as genetic diseases.

2. *A diagnostic index, being a list of referrals / cases seen by a centre or a consultant, organized on the basis of diagnostic information.* Such an index may be

used for research purposes, for audit, to arrange follow-up appointments, to re-contact families when a major advance has been made in understanding the condition in their family, or for a variety of other purposes. Again, such indices are clearly helpful to patient management in many contexts, and the principal issue relates to control of access to the register.

3. A register developed for gene mapping and genetic linkage analysis of families with a specific disorder; usually research-based. These registers are most likely to have been developed for genetic linkage analyses, either for research studies or for the clinical application of linkage results. A good example is the development of registers for research and counselling in relation to Duchenne muscular dystrophy [32]. Whether developed for research or counselling, many individuals never seen by the genetics staff may be on the register. These individuals will often be included in a purely formal sense because of their identity as a parent, or a healthy (unaffected) brother or sister; the computer program used for the linkage calculations may only function if, for example, every person at risk of the family's genetic condition is shown to have two parents. The parents need not be named, and may be long dead, but they must be listed. In this formal sense, individuals may be present on the register without having given their consent; they are not on the register as specific individuals but rather in a purely formal role as "Father (or mother) of individual no. 14". These individuals may or may not have provided a blood sample for research or for diagnostic purposes in relation to other family members, but many of them probably never gave explicit consent to be on a genetic register. They are included because the linkage studies would be impossible without their formal presence in the database.

This third type of register usually contains laboratory data (molecular genetic test results) and may contain some clinical information and personal identifying details (names, dates of birth . . .). These registers may still be maintained for those diseases where linkage analysis is still in process, but will have been superceded and often neglected where research has moved beyond linkage analysis.

The need for security in relation to such databases is obvious. The fact that individuals may be included in the register without their knowledge is not an ethical problem when their inclusion is purely formal. If they have been included in the genetic testing, however, and their personal results are in the database then they should clearly have given consent to having the genetic investigations performed. If that consent has been given, then it is not unreasonable to infer that consent was intended for the results of the tests to be stored for analysis. As discussed above, the importance of separating research results from standard medical records cannot be overemphasized.

4. A disease-specific register of families (or of individuals within families) at risk of late-onset disorders such as Huntington's disease. This type of register is very different. It functions as a list of individuals who are at risk of developing a genetic disorder in the future. Through such a register [33], those at-risk individuals in a family who want to be involved with the genetics centre can be contacted at regular intervals, often annually. This allows the genetics centre to update information about the family and to find out if any family members wish to have a clinic appointment to discuss the family's condition, or to be checked over for signs of the disorder, or to discuss the various options for reproduction (including prenatal diagnosis) or for predictive testing.

The nature of such registers varies within and between different genetics centres. Some registers may function passively, as a way of reviewing what contact there has been with the relevant individuals. Others, in contrast, function as active-contact registers, with genetics staff making annual contact with each individual by phone or letter, offering a clinic appointment or perhaps a visit to the family home, but only arranging a clinic appointment when that is requested.

The differences in policy between units depend upon staffing patterns and resources, as well as upon different opinions about what constitutes desirable practice. There are also differences as to what is appropriate in particular disease contexts.

5. A disease-specific register of individuals in families with a genetic disorder, for whom regular surveillance for complications may be appropriate. Regular surveillance of affected (or possibly affected) individuals for the early detection of tumours is appropriate in several of the familial cancer disorders, such as polyposis coli or von Hippel–Lindau disease [34–36]. This would often be co-ordinated through a regional genetics centre. A similar system of surveillance is also often appropriate for individuals or families with a firm diagnosis of Marfan syndrome, and there are other such conditions where a genetic register helps to ensure that at-risk individuals receive the best management available.

Individuals on the register will have expressed their interest in being kept under regular review. As long as access to the register is secure, few ethical problems will arise unless the register is used coercively in an attempt to impose surveillance on those who do not wish to be kept under review (and that is most unlikely).

6. A disease-specific register of possible carriers of X-linked, autosomal recessive and chromosomal conditions (i.e. registers of those who may be at risk of having children affected by a specific genetic disorder). Some centres have registers for a variety of different diseases organized in the same way as described in point 4 above. The purpose is generally two-fold:

(i) to offer '*cascade testing*' to identify those at risk of having affected children where carrier testing is technically straightforward (as with cystic fibrosis, haemoglobin disorders, familial chromosome rearrangements, etc.); and

(ii) to keep in contact with families for whom *the reproductive issues are complex and changing, or where young children can be offered genetic testing when they are older*. This allows family members to be given up-to-date information about their reproductive situation and ensures that the genetics unit has correct information about family structure and advance notice of likely requests for prenatal diagnosis. As children who might be carriers enter their reproductive years, the existence of a genetic register allows genetic counselling to be made available at a time appropriate to each family's circumstances (see Chapter 2).

There will be a continuing need for such registers as children who are or may be carriers grow older and as the methods of identifying carriers improve with the continuing progress in molecular genetic diagnostics.

General comments on genetic registers

For dealing with many diseases, the default state (i.e. no register) would be grossly inefficient. It would entail treating each individual in a family as independent, and waiting for separate referrals before seeing each family member. Each episode of counselling would then be regarded as self-contained, and a re-referral would be needed if an individual wished to be reassessed after a year or two. Such a cumbersome system would have many disadvantages, especially when clinical genetics should be aiming to provide a truly family-based service, following a number of individuals within a family over the course of years. The absence of genetic registers would then be very unsatisfactory for all involved, and would make it difficult for clinical geneticists to discharge their ongoing duty of care to individuals and families. A duty of care does not terminate ethically or legally just because a contracting system regards the episode of care as closed. The absurdities of a restrictive system of contracting in health care become very clear in a specialty such as clinical genetics in which individuals may live under the cloud of risk for many years before they develop any signs of disease, and in which a single referral may bring many members of a family into contact with the genetics centre. Genetic registers, then, are essential to good clinical practice.

At the same time, the existence of such registers could lead to problems of confidentiality or negligence (if those on the register are given suboptimal care or if the information contained in them is inaccurate), and many centres have not tackled the issue of consent, particularly for long-established registers. The origin of many registers as research tools that have now been superceded raises the question of whether such defunct registers should perhaps be erased.

Another serious and unresolved issue relating to 'genetic registers' is to determine where the clinical responsibility for their operation is located. If the point of a register is to maintain contact with families so that family members are offered regular surveillance for complications of a genetic disorder or updated information on reproductive risks, then is it the individual clinician who has ongoing responsibility for the extended family (even though s/he may move to another city), or the department, or the hospital, or the purchaser of health services who carries the responsibility of ensuring that the register performs its functions? This has been discussed by Pelias [37], although no consensus has yet emerged. If the purchaser fails to provide adequate funds for the register to operate effectively, then how liable is the clinician for any errors made? Perhaps a definition and description of 'genetic register' should be built into contracts between genetics departments and health care purchasers, with specified standards of operation.

What needs to be emphasized is the great variety of different entities that are now referred to in common medical parlance as 'genetic registers'. Very few such registers in fact function in the same way, reflecting their independent origins and their subsequent evolutionary paths, which have often depended upon technical developments in genetic testing. In any useful description of a genetic register, it is essential to include its underlying purpose as well as a description of what it is physically (a card index, a book, a computer file, . . .) and of how it is used in daily practice. It is also necessary to identify who carries responsibility for the register's successful operation.

CONCLUSIONS

Serious challenges to respect for confidentiality do arise in clinical genetics practice. These challenges should be acknowledged and confronted, and our respect for the confidentiality of personal genetic information should be reaffirmed. Any weakening of this respect will generate many more problems than it could possibly solve. Careful attention to several aspects of clinical and laboratory practice is essential if genetic services are not to jeopardize the genetic privacy of their clients. Genetic registers are not in themselves a threat to confidentiality and are required for good clinical practice, although access to the information in these registers must be carefully controlled. It will be helpful for the nature and purpose of registers to be kept under review.

REFERENCES

1. McLean SAM (1994) Mapping the human genome – friend or foe? *Soc. Sci. Med.* **39**: 1221–1227.
2. Reilly P (1992) ASHG statement on genetics and privacy: testimony to United States Congress. *Am. J. Hum. Genet.* **50**: 640–642.
3. Light DW (1992) The practice and ethics of risk-related health insurance. *J. Am. Med. Ass.* **267**: 2503–2508.
4. Harper PS (1993) Insurance and genetic testing. *Lancet*, **341**: 224–227.
5. Billings PR, Kohn MA, de Cuevas M, Beckwith J, Alper JS, Natowicz MR (1992) Discrimination as a consequence of genetic testing. *Am. J. Hum. Genet.* **50**: 476–482.

6. Ostrer H, Allen W, Crandall LA, Moseley RE, Dewar MA, Nye D, McCrary SV (1993) Insurance and genetic testing: where are we now? *Am. J. Hum. Genet.* **52**: 565–577.
7. Geller LN, Alper JS, Billings PR, Barash CI, Beckwith J, Natowicz MR (1996) Individual, family and societal dimensions of genetic discrimination: a case study analysis. *Sci. Eng. Ethics*, **2**: 71–88.
8. Nuffield Council on Bioethics (1993) *Genetic Screening – Ethical Issues.* Nuffield Council on Bioethics, London.
9. Lappe M (1988) ethical issues in genetic screening for susceptibility to chronic lung disease. *J. Occup. Med.* **30**: 493–501.
10. Rothenberg K, Fuller B, Rothstein M *et al.* (1997) Genetic information and the workplace: legislative approaches and policy challenges. *Science*, **275**: 1755–1757.
11. Mehlman MJ, Kodish ED, Whitehouse P, Zinn AB, Sollitto S, Berger J, Chiao EJ, Dosick MS, Cassidy SB (1996) The need for anonymous genetic counselling and testing. *Am. J. Hum. Genet.* **58**: 393–397.
12. Uhlmann WR, Ginsburg D, Gelehrter TD, Nicholson J, Petty EM (1996) Questioning the need for anonymous genetic counseling and testing. *Am. J. Hum. Genet.* **59**: 968–970.
13. Earley CL, Strong LC (1995) Certificates of confidentiality: a valuable tool for protecting genetic data. *Am. J. Hum. Genet.* **57**: 727–731.
14. McEwen JE, Reilly PR (1992) State legislative efforts to regulate the use and potential misuse of genetic information. *Am. J. Hum. Genet.* **51**: 637–647.
15. Scheck B (1994) DNA data banking: a cautionary tale. *Am. J. Hum. Genet.* **54**: 931–933.
16. Suslak L, Price DM, Desposito F (1985) Transmitting balanced translocation carrier information within families: a follow-up study. *Am. J. Med. Genet.* **20**: 227–232.
17. Wolff G, Back E, Arleth S, Rapp-Korner U (1989) Genetic counselling in families with inherited balanced translocations: experience with 36 families. *Clin. Genet.* **35**: 404–416.
18. Schaap T (1995) Confidentiality in counseling for X-linked conditions. *Clin. Genet.* **47**: 155–157.
19. Royal College of Physicians Committees on Ethical Issues in Medicine and Clinical Genetics (1991) *Ethical Issues in Clinical Genetics.* Royal College of Physicians of London, London.
20. Clarke A (1994) Genetic screening: a response to Nuffield. *Bull. Med. Ethics*, **940**: 13–21.
21. Boddington P (1994) Confidentiality in genetic counselling. In: *Genetic Counselling: Practice and Principles* (ed. A. Clarke), pp. 223–240. Routledge, London.
22. House of Commons Science and Technology Committee (1995) *Third Report. Human Genetics: the Science and its Consequences*, Vol. 1, *Report and Minutes of Proceedings*, paras 227–228. HMSO, London.
23. Manjoney DM, McKegney FP (1978) Individual and family coping with polycystic kidney disease: the harvest of denial. *Int. J. Psychiat. Med.* **9**: 19–31.
24. Shaw MW (1987) Testing for the Huntington gene: a right to know, a right not to know, or a duty to know. *Am. J. Med. Genet.* **26**: 243–246.
25. Pelias MZ (1992) The duty to disclose to relatives in medical genetics: response to Dr Hecht. *Am. J. Med. Genet.* **42**: 759–760.

26. British Medical Association (1992) *Medicine Betrayed: the Participation of Doctors in Human Rights Abuses*. Zed Books in association with the BMA: London and New Jersey.
27. Chadwick R (1993) What counts as success in genetic counselling? *J. Med. Ethics*, **19**: 43–46.
28. Harper PS (1993) Research samples from families with genetic diseases: a proposed code of conduct. *Br. Med. J.* **306**: 1391–1394.
29. ACMG Statement (1995) Statement on storage and use of genetic materials. *Am. J. Hum. Genet.* **57**: 1499–1500.
30. Clayton EW, Steinberg KK, Khoury MJ, Thomson E, Andrews L, Kahn MJE, Kopelman LM, Weiss JO (1995) Informed consent for genetic research on stored tissue samples. *J. Am. Med. Ass.* **274**: 1786–1792.
31. Harper PS (1992) Huntington disease and the abuse of genetics. *Am. J. Hum. Genet.* **50**: 460–464.
32. Read AP, Kerzin-Storrar L, Mountford RC, Elles RG, Harris R (1986) A register-based system for gene tracking in Duchenne muscular dystrophy. *J. Med. Genet.* **24**: 84–87.
33. Harper PS, Tyler A, Smith S, Jones P, Newcombe RG, McBroom V (1982) A genetic register for Huntington's Chorea in South Wales. *J. Med. Genet.* **19**: 241–245.
34. Littler M, Harper PS (1989) A regional register for inherited cancers. *Br. Med. J.* **298:** 1689–1691.
35. Burn J, Chapman P, Delhanty J *et al.* (1991) The UK Northern Region genetic register for familial adenomatous polyposis colui: use of age of onset, congenital hypertrophy of the retinal pigment epithelium and DNA markers in risk calculations. *J. Med. Genet.* **28**: 289–296.
36. Maddock IR, Moran A, Maher ER *et al.* (1996) A genetic register for von Hippel–Lindau disease. *J. Med. Genet.* **33**: 120–127.
37. Pelias MZ (1991) Duty to disclose in medical genetics: a legal perspective. *Am. J. Med. Genet.* **39**: 347–354.

Chapter

12

OUTCOMES AND PROCESS IN GENETIC COUNSELLING

A.J. Clarke

INTRODUCTION

The evaluation of medical services on the basis of the outcomes of clinical practice is widely recognized as essential [1], both as a research activity and as audit of practice. For many branches of medicine, where clinical practice is essentially therapeutic, the measurement of outcomes will be straightforward – in principle, if not always in practice. Achieving a diagnosis will be valued in so far as it makes a difference to therapy and the outcome of case management. Assessing the physical symptoms suffered by the patient will be the essential means of evaluating the outcomes of the service provided.

Clinical genetic practice is not generally involved with therapy. Our 'patients' are generally healthy. Rather than treatments, our service provides information, explanation and options; the options relate to reproduction, predictive genetic tests and health surveillance. Apart from the surveillance for complications of genetic disease, these activities would not be expected to alter the physical condition of our clients. We dispense words, not tablets.

How, then, should the evaluation of genetic counselling be conducted? Should we define the measure of our success by how well our clients recall the information that we have passed on to them? Or with how genetic counselling has altered their reproductive plans? Or how they have in practice acted on the basis of such information? This chapter considers how to evaluate the genetic counselling of individuals and families; it does not examine the rather different contexts of population genetic screening or of the surveillance for treatable complications of genetic disease.

This chapter is based on a paper by A.J. Clarke, E. Parsons and A. Williams (1996) in *Clinical Genetics*, Vol. 50, pp. 462–469. © 1996 Munksgaard International Publishers Ltd., Copenhagen, Denmark.

INFORMATION AS OUTCOME MEASURE

Studies of the recall of information given in genetic counselling generally indicate that the information is provided in an effective manner. Thus, Somer *et al.* [2] report that 80% of 791 families counselled in Finland had an adequate knowledge of the mode of inheritance, and 74% of the recurrence risk. Kessler has reviewed other studies in which pre-counselling and post-counselling knowledge of reproductive risk and of diagnosis are compared; genetic counselling appears to be effective in increasing knowledge [3,4]. The important study by Sorenson and others, however, has shown that there is a great potential for improving the effectiveness of information giving [5]. Knowledge may not be retained so well after genetic testing carried out in a population carrier screening programme, such as for cystic fibrosis (CF), in which there may be less emphasis upon individual counselling [6,7]. High levels of long-term recall of information were found, however, after an antenatal carrier screening programme for the haemoglobinopathies. This may have resulted from the provision of pre-test information in a novel or more stimulating form (by video) or because the greater anxiety present in an antenatal context reinforced the memory [8,9].

Although counsellees do recall many items of information correctly after counselling, it would be difficult to use this as a measure of the effectiveness of the service. This would entail pre-judging the purpose of genetic counselling as understood by the clients, who may not value specific items of information as highly as professionals. Furthermore, it has long been recognized that risk information is interpreted by counsellees rather than simply being recalled as neutral facts [10,11]. Clients tend to view risks in binary form – as either destined to happen or not to happen – and process the numerical information in ways that are heavily influenced by the anticipated social consequences of their reproductive decisions [12,13]. Similarly, carriers of a serious X-linked disorder experience this fact in different ways at different life stages, depending upon the relevance of the information at the time; what is retained is the personal meaning of the information they have been given [14]. It is also known that the way in which risks are presented (framed) can influence the interpretation of recurrence risk figures [15]. The evaluation of a risk as 'high', 'moderate' or 'low' may also differ greatly between counsellee and counsellor, depending in part upon the numerical risk of recurrence and in part upon other factors such as the perceived burden of the disorder [16]. Clients often interpret risk figures more optimistically (as lower) than do counsellors, but a variety of individual and family factors modify these interpretations [17]. There are simply too many subjective and variable factors involved for a service to be evaluated by the recall of risk figures or other items of information by its clients. There may well be occasions on which such recall could be assessed as part of a research study, but not as a measure of the effectiveness of a regular service.

REPRODUCTIVE PLANS AS OUTCOME MEASURE

It is possible to examine the reproductive intentions of genetic counselling clients before and after counselling. In one study by Wertz and Sorenson [18], 43.5% of 628 counsellees indicated that their reproductive plans were influenced by the genetic counselling, although 52.7% of those whose plans were reported to have been influenced by counselling held the same plans as they held before the counselling. Those who had come to counselling to get information for making a decision, those who discussed their decision in depth with the counsellor, and those with higher education, were more likely to report that the counselling influenced their plans. It is not entirely clear, however, in what sense clients were influenced by the counselling if their reproductive plans did not alter, and reproductive uncertainty will often not be resolved by genetic counselling because so many of the factors that contribute to a decision are social rather than clinical [19].

One important factor that certainly does influence reproductive plans is the availability of prenatal diagnosis. In the absence of prenatal diagnosis for a disorder, reproductive plans correlate with the perceived burden of the condition rather than the statistical risk of recurrence [20]. In counselling couples at risk of neural tube defect, the advent of prenatal diagnosis had a major impact, increasing the number willing to embark upon a pregnancy [21]. More recent studies confirm that the availability of prenatal diagnosis is important in helping families at high risk (>15%) of a child with a severe disorder in deciding to have further children [22]. Couples who already had an affected child were less likely to try again in the absence of prenatal diagnosis. Other important factors were the desire to have children and personal experience of a close relative affected by the disorder.

Could the influence on reproductive plans be used as an outcome measure in genetic counselling? There would be many problems in using it in this way. While this may be a valid topic for research, it would not be helpful as a measure of outcome for a clinical service. First, as with recall of information, to use reproductive plans as an outcome measure for genetic counselling is to prejudge the service that is wanted by client families. Some of them will want information that will inform their reproductive planning, but there may be many other reasons for seeking a referral to a clinical geneticist. Second, there are far too many confounding factors that influence reproductive plans for this to be a measure of effectiveness of a clinical service. Third, the choice of timeframe in which to assess the impact on reproductive plans would be arbitrary. Finally, reproductive plans are essentially hypothetical, and would be a poor proxy for reproductive behaviour. Caution is clearly required in the interpretation of stated reproductive intentions, given the findings of studies comparing the hypothetical with the actual uptake of genetic tests in two different contexts – predictive testing for Huntington's disease (HD) and prenatal diagnosis for CF. In both cases, the actual uptake of testing was much lower than the uptake predicted from hypothetical attitude surveys: 10–15% vs. 60% for HD [23,24]; 17% vs. 51% for CF [25].

REPRODUCTIVE BEHAVIOUR AS OUTCOME MEASURE

Some attempts have been made to report the effects of genetic counselling on long-term reproductive behaviour. Genetic counselling can be followed by an increase in pregnancies, irrespective of reproductive risk, even when prenatal diagnosis is not available [20]. Reproductive risk and the availability of prenatal diagnosis have the expected effects, leading couples to be more or less likely to embark upon a pregnancy respectively [2]. Controlled studies of reproductive behaviour after genetic counselling, however, are almost totally lacking, and there are no reports of randomized trials examining post-counselling reproductive behaviour. One controlled study of 46 families with a child with Down syndrome, showed no influence of genetic counselling on reproductive behaviour over at least 18 months from the birth of the affected child; nine in each group of 23 families initiated at least one further pregnancy, and only three of the 18 couples utilized prenatal diagnosis (two in the counselled group, one non-counselled) [26].

There are very good reasons why this area has hardly been investigated. There are both practical and ethical problems in using reproductive behaviour as a measure of the effectiveness of genetic counselling. While it may be possible to gather outcome data for families seen because of concern during an ongoing pregnancy, there are serious practical difficulties in using reproductive outcomes in other contexts. These include the duration of follow-up required, and the inability to gather these data as a part of routine service operation. Also, it is clear that the relationship between reproductive risk and reproductive behaviour is complex, and mediated by many different social factors in addition to the medical 'facts' [27].

There are also serious ethical difficulties in any use of reproductive outcomes as a measure of the effectiveness of genetic counselling. While it is perfectly legitimate research to assess the impact of genetic counselling on reproductive plans and behaviour, it is altogether different to monitor the effectiveness of the service in this way. That would entail imposing a value judgement on particular behaviours, declaring the birth of an affected child as undesirable, and the uptake of prenatal diagnosis and the termination of an affected fetus (or the childlessness of the at-risk couple) as desirable. Such judgements are unacceptable, and must be left to individuals. "It is also important that those giving genetic counselling do not judge 'success' or 'failure' in terms of a particular outcome, and that they give support to families whatever their decision may be" [28].

For clinicians to know that their work would be evaluated in this frankly eugenic manner would impose intolerable pressures on them. Clinicians would feel anxious that their clients make the 'correct' reproductive decisions. Health service purchasers would doubtless regard the correct decisions as those that imposed least monetary cost on the health service, and so clinicians (obstetricians, perhaps, as well as geneticists) could be regarded as failing in their duties if too many of their patients had infants affected by disorders that could have been diagnosed prenatally. This would lead to a conflict between the clinicians' duties to their patients and to their employers, and would rapidly bring their service into well-deserved disrepute. Followed to the logical extreme, this policy would lead to a recommendation that pregnancy be terminated for any fetus likely to cost more in medical care and social benefits than he/she would pay in taxes over his/her lifetime. But why should genetics services be expected to generate financial savings in this way, when the provision of all other aspects of clinical medicine is accepted as costing money?

A more sophisticated means of evaluating genetics services relating to reproduction might be to count the 'informed reproductive choices made', regarding informed choice as itself a benefit [29]. It would be very difficult to count the number of reproductive choices made by those who choose to have no more children, or who choose to proceed with pregnancies without prenatal diagnosis or without further discussion with genetics services, so that it might be difficult in practice to distinguish this approach from the previous one. Furthermore, it is very difficult to regard choice *per se* as the goal of a clinical activity [30,31] – this would lead to a proliferation of progressively less useful tests. The numbers of births of healthy children who would not have been born without prenatal diagnostic testing could be another measure of reproductive outcome [32], although it would be difficult to obtain suitable control data to validate the approach. Such studies may be possible as research projects, but not as a routine means for the evaluation of clinical genetics services.

We have to accept that crude, numerical measures of success are simply not applicable in the field of reproductive genetics. We need instead to adopt measures of outcome that accord with the goals of genetic counselling. The prime goal could be formulated as the clarification of the client's reproductive risks and options, where this was wanted by the client or, more generally, promoting the client's understanding of their genetic situation. How does this relate to the goals of genetic counselling in general?

WHAT THEN IS GENETIC COUNSELLING?

At this point, we must consider what takes place in a genetic counselling clinic. Genetic counselling can be defined as, "An educational process that seeks to assist affected and/or at-risk individuals to understand the nature of the genetic disorder, its transmission and the options open to them in management and family planning" [33]. This definition makes it clear that genetic counselling is a process centred on the clients and their need to understand the condition in their family. Because of this focus on understanding and education, it is essential for the professionals to listen to their clients, to explore their present understanding and their questions and concerns. Reports of genetic counselling in practice show that this process of communication has often broken down, so that counsellors are frequently unaware of the issues that their clients want to discuss and these issues are therefore often not addressed [34,35]. It is interesting that the concerns of female clients attending genetic counselling on their own were more likely to be addressed by female than by male counsellors [36].

The ethos of genetic counselling, then, is for the clients to set the agenda, and the first element of genetic counselling must therefore be listening [37]. The clients' questions may focus on the diagnosis or prognosis of the condition being considered, or on reproductive risks and options, or on other issues. The structure of a genetic counselling service is clearly important in deciding whether or not the client's concerns will be heard, in turn deciding whether or how a client's concerns will be addressed.

The next element is the process of clinical assessment that allows the questions to be answered, followed by communicating the relevant information to the family. Then, for some clients, there is the facilitation of decisions (scenario-based decision counselling) that helps clients to consider the practical and emotional consequences for each of the different possible outcomes of any reproductive or testing decisions that confront them. Finally, there is the provision of ongoing support where this is appropriate.

This model of genetic counselling operates through the process of clinical service that we provide in Wales. After a referral is received, our first point of contact with the client is usually a visit to the family home by the local genetics nurse. At this visit, there is a two-way exchange of information. The nurse asks about the family's questions and concerns, and also gathers information about the condition or event in the family that has prompted the referral, and more generally about the family history. Who in the family has been affected? Are previous medical records available on other relevant family members? Will it be possible to obtain their consent for these records to be reviewed? Gathering this information in advance often reduces the length and number of clinic appointments required by the family. At the same time, the genetics nurse is able to explain the ethos of the service, often reassuring the family that we will not be telling them what course of action to take, and she will also explain the limitations of the service – for example, that we may not be able to arrive at a firm diagnosis to account for a child's developmental delay, or that the prognosis for an infant with a particular congenital anomaly may be uncertain. The role of the genetics nurse/associate/co-worker is particularly important in carrying out this pre-clinic exchange of information with the family or individual referred for genetic counselling, and is a major strength in the provision of genetic counselling services in many areas of Britain [38].

The next point of contact is usually a clinic visit (or a few visits) to see the nurse and clinical geneticist. The family's questions and concerns are elicited once more in the clinic – they may have changed since the preliminary discussion with the nurse. The clinician asks the necessary clinical questions, and performs the appropriate physical examination and investigations, so that the client's questions can be answered as far as possible. The clinician's conclusions are explained to the clients and discussed with them. If the clients are confronting important decisions, perhaps in relation to predictive testing, carrier status testing, or testing of an ongoing pregnancy, then the various options will be discussed, and the implications for the clients and their families of the possible test results, or of not having the test, will be explored. Finally, arrangements will be made for future contact where appropriate. This may involve contact with other members of the client's family, a planned prenatal diagnosis, the provision of support after a termination of pregnancy or for those living at risk of serious disease, or a review for diagnostic reassessment or for surveillance for early signs of complications of the disorder.

There are other aspects of clinical genetics, but they do not fit so well within this description of genetic counselling. Our role in diagnosing rare disorders may be related to our role in genetic counselling but it is distinct from it, and it too is difficult to assess in simple quantitative terms. Clinical genetics can also be taken to include the co-ordination of surveillance for early signs of potentially treatable complications of such conditions as familial cancers and Marfan syndrome. The evaluation of these aspects of clinical genetics fits much more readily into standard assessments of outcome, because they relate to physical health. The provision of population screening for some genetic conditions, (e.g. newborn screening for phenylketonuria and hypothyroidism, antenatal screening for Down syndrome, or carrier screening for recessive disorders), is also not a part of family-centred genetic counselling and may not be provided through Departments of Clinical Genetics, although such screening programmes are a part of clinical genetics in the broader sense.

SATISFACTION WITH GENETIC COUNSELLING

Could satisfaction with the process of genetic counselling, as outlined above, be used as an outcome measure? Counsellor satisfaction with counselling sessions has been reported to be very high [39], and appears to depend upon the perception that the counsellor has communicated knowledge successfully (preferably to better-educated clients) rather than to any meeting of the clients' needs. For satisfaction to be considered as an outcome of genetic counselling, it will be necessary to focus upon the satisfaction of clients, not counsellors. Few attempts to measure client satisfaction with genetic counselling have been reported, however, and a considerable problem could be dissatisfaction associated with the lack of available information or with unwelcome information provided in the course of genetic counselling [40,41] – blaming the messenger. Simple measures of client mood or contentment after the counselling will obviously be inadequate. Such measures as a short form of the Spielberger

State–Trait Anxiety Inventory and the General Health Questionnaire have been used to assess the psychological impact of population screening programmes, and have generally shown that subjects are stressed when given unwelcome information (e.g. that they are a carrier of CF) but that this distress is only moderate and is not sustained [42,43]. Even in that context, the "psychological consequences (of population carrier testing) are unclear" [44]. Such measures are simply not appropriate in the much more varied context of generic genetic counselling. The sharing of information in genetic counselling may often cause sustained and thoroughly appropriate distress, as when an adult, child or fetus is diagnosed as having a serious, degenerative disorder. What measure of satisfaction, or emotional well-being, could possibly be appropriate to the varying circumstances of a genetics clinic?

The possibility of using a sophisticated form of quality of life (QOL) evaluation for genetics services has been raised [45], but would involve such a simplification of complex outcomes that it would inevitably lead " . . . to unwise or unfair decisions" [46]. Multidimensional measures of QOL would certainly be more suitable than any single, global measure of QOL [47], but will still be inadequate to permit valid comparisons of health gains resulting from genetic counselling with those resulting from other areas of clinical medicine.

Shiloh *et al.* have devised a scale for the evaluation of instrumental, affective and procedural aspects of genetic counselling [40], but this evaluates the process of genetic counselling rather than clear endpoints or outcomes. Such measures will need to be refined, but hardly answer the question of what is achieved in genetic counselling. If no diagnosis is achieved for a child with serious disabilities, for example, the clinician may not be at fault (there may be no diagnosis to make) but the parents may well feel dissatisfied [48].

It is clear from a consideration of patient satisfaction surveys in general, not just in genetics, that attempts to assess global levels of satisfaction with a service will not be helpful in identifying either the strengths or the weaknesses of a service [49]. Instead, attention must be focused on specific aspects of performance wherever possible [50]. Few such studies have been conducted in genetic counselling. It has been shown that a routine visit by a genetics nurse to the family home after genetic counselling does not improve the client's recall of information [51]. It has also been shown that the use of structured scenarios in genetic counselling can be very helpful [52], and also a potentially powerful way of shaping decisions [53].

Given the difficulty in finding suitable outcome measures for most genetic counselling, and the problems with relying on satisfaction in a broad sense, it could be suggested that our achievements are not related to health as such, and should therefore not be supported by health service money. Indeed, if health service activity is to be defined as activities that improve simple measures of physical and mental well-being, such as the SF-36[54,55] and the General Health Questionnaire, then perhaps genetic counselling is outside the domain of the health services: it should not be competing for the same pot of funds as treatments for broken bones or acute pneumonia. While this definition of health service activity may be appropriate in many areas, however, it is unacceptably narrow in the context of genetic counselling [47,56].

QUALITY OF PROCESS IN GENETIC COUNSELLING

In response to the manifest inadequacy of outcome measures in genetic counselling, Kessler has suggested that we need to focus on the process of genetic counselling itself: "We probably have come almost as far as we can in terms of applying outcome methodology and it may be time now to focus on the counselling process itself. . . . Perhaps then we can assess where success is being achieved and where remedy is needed" [4].

Similar considerations apply to the evaluation of counselling and support services in other contexts such as primary health care and community paediatrics, where simple measures of well-being are also inadequate as measures of outcome [57,58]. In such contexts, where no medical interventions are prescribed and where the activity being evaluated is primarily a process of communication, it may be more appropriate to rely upon research evidence to decide what services to offer, and then to assess the quality of the process of the service provided, rather than continuing to grope towards elusive measures of outcome. Indeed, genetic counselling may be particularly suited to the purchasing of protocols of care rather than of activity as such [59].

Research is required into the process of genetic counselling, that will enable us to describe and evaluate different models of genetic counselling services. Purchasers will then be able to use the research evidence to decide which model of genetic counselling to select. We should be judged by our choice of process, and by how well we meet the appropriately specified service standards, and not by attempts to monitor reproductive or other outcomes in this delicate field. This argument is a form of special pleading, but it will be recognized as valid by, for example, those health economists who recognize that monetary valuation of prenatal screening is unsuitable, so that cost–benefit analyses of these services have to be excluded [60]. Once this is accepted more generally, the need for the monitoring of genetic counselling services to focus on quality issues and not outcomes will then be much more acceptable to purchasers.

A PROPOSAL

The inadequacy of relying on statements of satisfaction with a service made by naïve clients has been emphasized by Calnan in his development of a framework for the lay evaluation of health care [61]. If satisfaction is to be assessed, it would only be appropriate once the client is sufficiently informed to make a reasonable judgement about the quality of the service provided. For genetic services, clients may be in this position once the process of genetic counselling is completed.

This leads to the 'retrospective assessment of satisfaction' approach, which moves beyond the simplistic, quantitative measures of client satisfaction with health care so rightly criticized by Carr-Hill [56] and Williams [49]. This could generate a much more valid measure of clinical worth than a client satisfaction questionnaire administered or posted to clients soon after a clinic appointment.

The principal tool for evaluating genetic counselling, therefore, may have to be the confrontation of clients with three questions:

(i) Were you satisfied with the service provided at the time? Was the service efficient and convenient to use? – incorporating the dimensions of satisfaction of Shiloh *et al.* [40].
(ii) What changes have occurred in your expectations of the genetic counselling service since your referral? – a measure of the clients' education about genetic counselling and of their adjustment to their genetic situation.
(iii) With hindsight, how satisfied are you with your experience of the genetic counselling service? If you went through a diagnostic process, was this of any use or value, even if no diagnostic certainty was achieved? Did the service meet the expectations that you now think you 'should' have had at the time of your referral?

This approach acknowledges that there may be a mismatch between the client's initial expectations and what the service can provide, and that the client is only in a position to judge this when the episode of genetic counselling has been completed. This perspective is supported by Michie *et al.*, who suggested that it may be appropriate for genetic counsellors to alter the unrealistic or inappropriate expectations of their clients rather than try to meet them [62]. Their study of the expectations and satisfaction of clients with genetic counselling chose satisfaction with six dimensions of genetic counselling (information, explanation, reassurance, advice, help in making decisions and 'anything else') and the clients' levels of anxiety as its outcome measures. These dimensions of satisfaction are clearly not independent, however, and the meaning of some terms is ambiguous, so the findings of this study can only be regarded as giving a crude insight into the process of genetic counselling. Some of the dimensions of counselling are frankly unsuitable for use in assessing the quality of the process of counselling or its outcomes; measures of 'reassurance', for example, are clearly inappropriate for the reasons discussed above. It will be possible for future studies to build upon this work, however, and incorporate more sophisticated means of understanding, representing and evaluating what occurs within genetic counselling.

Adopting the retrospective evaluation of satisfaction approach to the evaluation of genetic counselling would enable evaluation to be an ongoing part of service activity, requiring few additional resources. The necessary initial data can be gathered as a routine by the genetics nurse at her first contact with each family, and the retrospective assessment of the service by each client could be incorporated into a standard questionnaire sent to each client at the end of an episode of genetic counselling. In addition to this continuing feedback from clients, it would also be necessary to carry out peer review of service quality and clinical audit, as in other units, and to contact the referring doctors to discover how satisfied they have been with the service provided to their patients.

Simpler measures of outcome may be superficially attractive but are neither methodologically nor ethically acceptable. While the type of evaluation proposed would provide some subjective outcome data, however, it must be recognized that it still fails to provide a global outcome measure for the entire process of genetic counselling. Such a measure, however, may be unattainable; attempts to devise such a measure could lead to inappropriate and even destructive, Procrustean efforts to reshape the activity of genetic counselling.

This discussion also touches on a broader debate about the role of qualitative methodologies in public health analyses. Current practice is to focus largely on quantitative methods, assuming that benefits exist only if they can be counted, while qualitative methods are also required if purchasing decisions in the (British) National Health Service are to do justice to the array of competing claims in certain supremely sensitive areas [49,63]. Clinical genetics services comprise one such area of clinical practice in which the need to incorporate qualitative assessments of process into the evaluation of services is pressing because the exclusive adoption of simply quantifiable outcomes may result in great harm to patients and families.

REFERENCES

1. Delamothe T (1994) Using outcomes research in clinical practice. *Br. Med. J.* **308**: 1583–1584.
2. Somer M, Mustonen H, Norio R (1988) Evaluation of genetic counselling: recall of information, post-counselling reproduction, and attitude of the counsellees. *Clin. Genet.* **34**: 352–365.
3. Kessler S (1989) Psychological aspects of genetic counselling VI. A critical review of the literature dealing with education and reproduction. *Am. J. Med. Genet.* **34**: 340–353.
4. Kessler S (1992) *Process Issues in Genetic Counselling*. Birth Defects: Original Article Series, vol. 28, pp. 1–10. Alan R. Liss, New York.
5. Sorenson JR, Scotch NA, Swazey JP (1981) *Reproductive Pasts, Reproductive Futures. Genetic Counselling and its Effectiveness*. Birth Defects: Original Article Series, vol. XVII. Alan R. Liss, New York.
6. Bekker H, Denniss G, Modell M, Bobrow M, Marteau T (1994) The impact of population based screening screening for carriers of cystic fibrosis. *J. Med. Genet.* **31**: 364–368.

7. Axworthy D, Brock DJH, Bobrow M, Marteau TM (1996) Psychological impact of population-based carrier testing for cystic fibrosis: 3-year follow-up. *Lancet*, **347**: 1443–1446.
8. Loader S, Sutera CJ, Walden M, Kozyra A, Rowley PT (1991) Prenatal screening for hemoglobinopathies. II. Evaluation of counselling. *Am. J. Hum. Genet.* **48**: 447–451.
9. Loader S, Sutera CJ, Segelman SG, Kozyra A, Rowley PT (1991) Prenatal hemoglobinopathy screening. IV. Women at risk for a child with a clinically significant hemoglobinopathy. *Am. J. Hum. Genet.* **49**: 1292–1299.
10. Pearn JH (1973) Patients' subjective interpretation of risks offered in genetic counselling. *J. Med. Genet.* **10**: 129–134.
11. Evers-Kiebooms G, Berghe van den H (1979) Impact of genetic counselling: a review of published follow-up studies. *Clin. Genet.* **15**: 465–474.
12. Lippman-Hand A, Fraser F Clarke (1979) Genetic counselling: provision and reception of information. *Am. J. Med. Genet.* **3**: 113–127.
13. Lippman-Hand A, Fraser F Clarke (1979) Genetic counselling: the post-counselling period: 1. Patients' perceptions of uncertainty. *Am. J. Med. Genet.* **4**: 51–71.
14. Parsons EP, Atkinson PA (1992) Lay construction of genetic risk. *Sociol. Hlth Illness*, **14**: 437–455.
15. Shiloh S, Sagi M (1989) Effect of framing on the perception of genetic recurrence risks. *Am. J. Med. Genet.* **33**: 130–135.
16. Emery AEH, Raeburn JA, Skinner R, Holloway S, Lewis M (1979) Prospective study of genetic counselling. *Br. Med. J.* **i**: 1253–1256.
17. Wertz DC, Sorenson JR, Heeren TC (1986) Clients' interpretation of risks provided in genetic counselling. *Am. J. Hum. Genet.* **39**: 253–264.
18. Wertz DC, Sorenson JR (1986) Client reactions to genetic counselling: self-reports of influence. *Clin. Genet.* **30**: 494–502.
19. Wertz DC, Sorenson JR, Heeren TC (1984) Genetic counselling and reproductive uncertainty. *Am. J. Med. Genet.* **18**: 79–88.
20. Sorenson JR, Scotch NA, Swazey JP, Wertz DC, Heeren TC (1987) Reproductive plans of genetic counselling clients not eligible for prenatal diagnosis. *Am. J. Med. Genet.* **28**: 345–352.
21. Morris J, Laurence KM (1976) The effectiveness of genetic counselling for neural-tube malformations. *Dev. Med. Child Neurol.* **18** (suppl. 37): 157–163.
22. Frets PG, Duivenvoorden HJ, Verhage F, Niermeijer MF, Berge SMM van de, Galjaard H (1990) Factors influencing the reproductive decision after genetic counselling. *Am. J. Med. Genet.* **35**: 496–502.
23. Tyler A, Ball D, Craufurd D (1992) Presymptomatic testing for Huntington's disease in the United Kingdom. *Br. Med. J.* **304**: 1593–1596.
24. Harper PS, Ball DM, Hayden MR (1993) Presymptomatic testing for Huntington's disease: a worldwide survey. *J. Med. Genet.* **30**: 1020–1022.
25. Jedlicka-Kohler I, Gotz M, Eichler I (1994) Utilization of prenatal diagnosis for cystic fibrosis over the past seven years. *Pediatrics*, **94**: 13–16.
26. Oetting LA, Steele MW (1982) A controlled retrospective follow-up study of the impact of genetic counselling on parental reproduction following the birth of a Down syndrome child. *Clin. Genet.* **21**: 7–13.
27. Parsons EP, Atkinson P (1993) Genetic risk and reproduction. *Sociol. Rev.* **41**: 679–706.

28. Harper PS (1988) *Practical Genetic Counselling*, 3rd Edn. Wright, London.
29. Modell B, Kuliev AM (1993) A scientific basis for cost–benefit analysis of genetics services. *Trends Genet.* **9**: 46–52.
30. Chadwick RF (1993) What counts as success in genetic counselling? *J. Med. Ethics*, **19**: 43–46.
31. Clarke A (1993) Response to: What counts as success in genetic counselling? *J. Med. Ethics*, **19**: 47–49.
32. Beech R, Rona RJ, Swan AV, Wilson OM, Mandalia S (1992) A methodology for simulating the impact of DNA-probe services on the outcomes of pregnancies. *Int. J. Tech. Assess. Hlth Care*, **8**: 539–545.
33. Kelly TE (1986) *Clinical Genetics and Genetic Counselling*. Year Book, Chicago.
34. Sorenson JR, Swazey JP, Scotch NA (1981) *Reproductive Pasts, Reproductive Futures. Genetic Counselling and its Effectiveness*. Birth Defects: Original Article Series, vol. XVII. Alan R. Liss, New York.
35. Wertz DC, Sorenson JR, Heeren TC (1988) Communication in health professional–lay encounters: how often does each party know what the other wants to discuss? In: *Information and Behaviour* (ed. B.D. Ruben) Vol. 2, pp. 329–342. Transaction Books, New Brunswick.
36. Zare N, Sorenson JR, Heeren T (1984) Sex of provider as a variable in effective genetic counselling. *Soc. Sci. Med.* **19**: 671–675.
37. Clarke A (1994) *Introduction to Genetic Counselling: Practice and Principles*. Routledge, London.
38. Farnish S (1988) A developing role in genetic counselling. *J. Med. Genet.* **25**: 392–395.
39. Wertz DC, Sorenson JR, Heeren TC (1988) 'Can't get no (dis)satisfaction'. Professional satisfaction with professional–client encounters. *Work Occup.* **15(1)**: 36–54.
40. Shiloh S, Avdor O, Goodman RM (1990) Satisfaction with genetic counselling: dimensions and measurement. *Am. J. Med. Genet.* **37**: 522–529.
41. Mushlin AI, Mooney C, Grow V, Phelps CE (1994) The value of diagnostic information to patients with suspected multiple sclerosis. *Arch. Neurol.* **51**: 67–72.
42. Bekker H, Modell M, Denniss G, Silver A, Mathew C, Bobrow M, Marteau T (1993) Uptake of cystic fibrosis testing in primary care: supply push or demand pull? *Br. Med. J.* **306**: 1584–1586.
43. Mennie M, Compton ME, Gilfillan A, Liston WA, Pullen I, Whyte DA, Brock DJH (1993) Prenatal screening for cystic fibrosis: psychological effects on carriers and their partners. *J. Med. Genet.* **30**: 543–548.
44. Marteau T (1994) Psychological consequences are unclear. *Br. Med. J.* **309**: 1429–1430.
45. Johal S (1995) Quality of life: an evaluation alternative for genetic services. *Psychol. Update*, in press.
46. Spiegelhalter DJ, Gore SM, Fotzpatrick R, Fletcher AE, Jones DR, Cox DJ (1992) Quality of life measures in health care. III. Resource allocation. *Br. Med. J.* **305**: 1205–1209.
47. Jenkins C David (1992) Assessment of outcomes of health intervention. *Soc. Sci. Med.* **35**: 367–375.
48. Abramovsky I, Godmilow L, Hirschhorn K, Smith H (1980) Analysis of a follow-up study of genetic counselling. *Clin. Genet.* **17**: 1–12.

49. Williams W (1994) Patient satisfaction: a valid concept? *Soc. Sci. Med.* **38**: 509–516.
50. Bruster S, Jarman B, Bosanquet N, Weston D, Erens R, Delbanco TL (1994) National survey of hospital patients. *Br. Med. J.* **309**: 1542–1549.
51. Curtis D, Johnson M, Blank CE (1988) An evaluation of reinforcement of genetic counselling on the consultant. *Clin. Genet.* **33**: 270–276.
52. Arnold JR, Winsor EJT (1984) The use of structured scenarios in genetic counselling. *Clin. Genet.* **25**: 485–490.
53. Huys J, Evers-Kiebooms G, d'Ydewalle G (1992) *Decision Making in the Context of Genetic Risk: the Use of Scenarios*. Birth Defects: Original Article Series, vol. 28, pp. 17–20. Alan R. Liss, New York.
54. Brazier JE, Harper R, Jones NMB, O'Cathain A, Thomas KJ, Usherwood T, Westlake L (1992) Validating the SF-36 health survey questionnaire: new outcome measure for primary care. *Br. Med. J.* **305**: 160–164.
55. Garratt AM, Ruta DA, Abdalla MI, Buckingham JK, Russell IT (1993) The SF 36 health survey questionnaire: an outcome measure suitable for routine use within the NHS? *Br. Med. J.* **306**: 1440–1444.
56. Carr-Hill RA (1992) The measurement of patient satisfaction. *J. Publ. Hlth Med.* **14**: 236–249.
57. Hazzard AJ (1995) Measuring outcome in counselling: a brief exploration of the issues. *Br. J. Gen. Pract.* **45**: 118–119.
58. Milner J, Bungay C, Jellinek D, Hall DMB (1996) Needs of disabled children and their families. *Arch. Dis. Childhood*, **75**: 399–404.
59. Sheldon TA, Borowitz M (1993) Changing the measure of quality in the NHS: from purchasing activity to purchasing protocols. *Qual. Hlth Care*, **2**: 149–150.
60. Phin N (1990) *Can Economics be Applied to Prenatal Screening?* Discussion Paper 74, Centre for Health Economics, University of York, York.
61. Calnan M (1988) Towards a conceptual framework of lay evaluation of health care. *Soc. Sci. Med.* **27**: 927–933.
62. Michie S, Marteau TM, Bobrow M (1997) Genetic counselling: the psychological impact of meeting patients' expectations. *J. Med. Genet.* **34**: 237–241.
63. Baum FG (1995) Researching public health: behind the qualitative–quantitative methodological debate. *Soc. Sci. Med.* **40**: 459–468.

Chapter

13

THE PROCESS OF GENETIC COUNSELLING

Beyond non-directiveness

A.J. Clarke

INTRODUCTION

The primary goal of genetic counselling may be seen as the provision of information, as in this definition:

> Genetic counselling is the communication of information and advice about inherited conditions .... [1].

The emphasis here is on the technical aspects of genetics. This reflects the practice of many clinical geneticists, but it contrasts with the broader definition of genetic counselling given in 1974 by Fraser, in which it is described as:

> a communication process which deals with the human problems associated with the occurrence, or risk of occurrence, of a genetic disorder in a family. This process involves an attempt by one or more appropriately trained persons to help the individual or the family to:
> (i) comprehend the medical facts, including the diagnosis, the probable course of the disorder and the available management;
> (ii) appreciate the way heredity contributes to the disorder and the risk of recurrence in specified relatives;
> (iii) understand the options for dealing with the risk of recurrence;
> (iv) choose the course of action which seems appropriate to them in view of their risk and their family goals and act in accordance with that decision;
> (v) make the best possible adjustment to the disorder in an affected family member and/or to the risk of recurrence of that disorder [2].

This definition moves far beyond establishing a diagnosis and providing information, to encompass helping the family to make the decisions they find most appropriate and to make the best adjustment to their situation. These areas of decision-making and personal adjustment are where the counselling component of genetic counselling is so important. Clinical geneticists do attend to these topics, but often with less confidence and less professionalism than when they deal with the more technical and impersonal aspects of their work; I am speaking for myself here, but also I think for a good number of colleagues.

One particular area in clinical genetics in which the decision-making process of clients is regularly assessed and promoted in a structured fashion is in the context of predictive testing for late-onset disorders such as Huntington's disease (see Chapter 3 for a discussion of this). Counselling in relation to prenatal diagnostic decisions is often not approached in such a systematic way, especially in routine antenatal clinics. This is probably because of the rather haphazard way in which these services evolved, in contrast with the carefully planned and co-ordinated development of protocols for predictive testing in families with Huntington's disease.

Assisting clients with decision-making and with personal adjustment to their situation are aspects of genetic counselling that are not reflected in all definitions, and which are relatively poorly attended to in many training programmes for clinical geneticists. Only a few clinical geneticists have much formal training in psychotherapy or 'generic' counselling (i.e. counselling in a broad sense). Those with such experience are perhaps able to explore these issues of decision-making and personal adjustment more fully, and so can the non-medical North American genetic associates, who are trained in counselling as well as in genetics.

NON-DIRECTIVENESS

It is clear from Fraser's definition of genetic counselling that the role of the clinical geneticist or genetic counsellor in relation to the making of decisions is envisaged as helping clients to make the decision that seems most appropriate to them as individuals. The role of genetic counselling is not to lead clients to make particular decisions or choices (those preferred or recommended by the clinician, the health service or by society) but to help them to make the best decisions for themselves and their families as judged from their own perspectives. This facilitative approach (not prejudging what decision should be made from anyone else's perspective) can be described as being non-directive. The term has been borrowed from the non-directiveness advocated by Carl Rogers in his writings about client-centred therapy.

If genetic counselling is indeed to be non-directive, then there are important implications for any evaluation of its outcomes; these are discussed in Chapter 12. That chapter develops the argument, put forward in a paper entitled "Genetics, ethics and audit" [3], that clinical geneticists could be led towards imposing reproductive decisions on clients because of administrative pressure aimed at maximizing the financial 'savings' resulting from genetic screening and counselling. I argued that genetic counsellors should explicitly state that their goal is not to guide their clients to particular decisions but to help them to make their own, and that they should reject the use of crude measures of outcome (such as the number of terminations of pregnancy for a given condition). I thought that this would clarify uncertainty about our role in the minds of our clients and would be sufficient to ensure that we practised in a truly non-directive manner. The paper argued for the very clear separation of genetic counselling practice from public health policies and goals aimed at a whole population; the importance of this distinction has been emphasized by Harper in this book and elsewhere [4].

'STRUCTURAL' DIRECTIVENESS

On further reflection, I came to realize that pressure applied to individuals and families in relation to their reproductive decisions need not come consciously from misguided professionals (although that certainly can and does happen) but could also arise out of the very context of genetic counselling, without any conscious intention by professionals to coerce or manipulate their clients. This argument was put forward in another paper, entitled "Is non-directive genetic counselling possible?" [5], which maintained that genetic counselling is provided in such a context that it is bound to be perceived and experienced as leading clients towards specific outcomes. Even if counselling appears superficially to be conducted in a non-directive manner, it may function implicitly as directive [6]. Merely making prenatal diagnosis and screening available may itself lead to our clients feeling that we are promoting such tests. This entails a paradox, in that if we do not make such tests available then we are clearly limiting the choices of our potential clients, but in a different way.

These difficulties do not arise as a consequence of the personal failure of individual clinicians. The fact that prenatal diagnostic and screening tests are available imposes a burden of responsibility on every couple embarking upon a pregnancy, whether at increased or standard risk of having a child with a serious problem. This burden of responsibility will lead some individuals to feel pushed, against their wishes, to accept whatever offers of prenatal testing are made. The imposition of such feelings is not the 'fault' of individual clinicians, midwives or counsellors; rather, they arise from the level of technological development of contemporary Western society. The existence of such tests leads not only to the burden of responsibility and the sense of obligation to undergo testing, but leads in other members of society to a sense of the mother being to blame for the birth of a child with a 'preventable' condition such as Down syndrome [3,7]. In genetic counselling we can try to make these feelings more explicit, and so help individuals to understand the pressures they experience and then come to considered decisions about what to do; we cannot turn back the clock and pretend that such tests do not exist.

We can avoid reinforcing these pressures to participate in reproductive genetic testing, but we cannot remove them. Despite the wishes and intentions of many individual professionals, we operate in a system in which genetic counselling and screening services are perceived as a way of reducing the number of infants born with genetic disorders. This inappropriate understanding of clinical genetics will be reinforced unless we explicitly reject it; otherwise, our silence will allow this focus on endpoints and outcomes to dominate and distort our clinical work to the detriment of our patients and of society. Therefore, we must take care not to use such cost–benefit considerations to justify health service support for genetics services; these considerations, however, may nevertheless be instrumental in persuading health authorities to do so. This could undermine, or even demolish, our claim to be non-directive.

The publication of the 1991 paper [5] stirred up a useful debate about genetic counselling and its goals and consequences. While my concerns about *structural* rather than *individual* directiveness were echoed by a number of colleagues, others chose to take offence at my critical tone. The sensitivity exposed in the course of this debate was revealing. Such reactivity suggested the dim apprehension of an unpalatable truth. It is of course tempting to cling to slogans, such as 'genetic counselling is non-directive', when the alternative is to feel discomfort about one's professional role. My own view is that genetic counselling inevitably entails tensions between conflicting values. Perhaps the principal tension embedded within genetic counselling, at least in relation to reproductive issues, is that between respect for those affected by genetic conditions on the one hand and the promotion of informed reproductive decisions on the other hand. I am sure that it is much better to be aware of these tensions, and to seek an accommodation between them, than to suppress the awareness.

WHY NON-DIRECTIVENESS?

Why is the temptation to cling to non-directiveness so strong? Why do clinical geneticists feel the need to be protected by adherence to this doctrine? There are at least four answers.

First is the great respect in which autonomy is held. Autonomy is now the dominant principle of medical ethics, at least in the Western world. Any unsolicited professional intrusion, particularly in this intensely personal realm of reproductive decisions, would be regarded as an impertinence.

Secondly, there are good reasons for our profession wishing to dissociate itself from its recent past – from the abuses of human genetics in both Europe and North America over the first half of this century[a]. If a concern for genetic

---

[a]These abuses are often summed up in the emotive word 'eugenics', but it must be remembered that eugenics and human genetics were closely associated for many years as worthy and respected disciplines, with 'eugenics' really meaning 'applied human genetics'; it is only more recently that eugenics has come to have overwhelmingly negative connotations (see Section IV).

aspects of 'the public health' is driven by a society (or a part of society) that is unwilling to meet the costs of compassionate caring and support for those with disabilities and disease, then the success of genetic services will be judged simply on the basis of comparative costs of screening and pregnancy termination versus the cost of caring for affected individuals. If pushed to its logical extreme, such 'public health genetics' leads to horrors such as the Nazi programme of racial hygiene, which in turn led to compulsory sterilization and subsequently to medicalized murder. It is very clear why we would all wish to dissociate ourselves from such practices, but some contemporary genetic screening programmes are justified in essentially these terms.

Another reason for the wish to claim adherence to non-directiveness is the emotional distance that it lends to our professional work. It helps us to be less emotionally involved in the intimate personal lives of our clients, and this may be a very useful and healthy defence against over-involvement, as long as this role is recognized. It protects us from our feelings when clients make decisions with which we disagree; it helps us to cope with the cognitive dissonance that results from our supporting our clients in their decisions despite our (hopefully concealed) discomfort, distress or concern at their decisions and actions [8]. Finally, this sense of distance (the clear removal of the genetic counsellor from personal responsibility for their clients' decisions) may also reassure us that we will not be held legally accountable for those – their – decisions.

NON-DIRECTIVENESS IN PRACTICE?

Observations of genetic counselling practice and of talk amongst clinical geneticists illustrate how frequently we refer to the ideal of non-directiveness, trying not to influence our clients unduly and trying not to let our own values or opinions show through our professional masks [9]. In this way, non-directiveness is a rhetorical cornerstone of our professional identities.

Fletcher, Wertz and their colleagues have conducted a number of surveys of opinion among clinical geneticists, which demonstrate the widespread support for a non-directive ethos in genetic counselling practice [10–12]; however, there may be some limits to the willingness of geneticists to support every decision made by their clients [12]. A willingness to be somewhat directive in practice has also been reported in several other attitude surveys, such as those conducted by Marteau and her colleagues. In relation to counselling following the diagnosis of a fetal abnormality, it seems that relevant factors include the condition diagnosed, the nationality of the genetic counsellor and the professional background of the professional. In one study, geneticists in Germany and Portugal appeared to be more directive than those from Britain, with those from Germany reporting a greater willingness to advise continuation of a pregnancy and those from Portugal being more likely to advise termination of pregnancy [13]. Furthermore, it seems that obstetricians are more likely to be directive (advising termination of a pregnancy) than are clinical geneticists, who are more likely to be directive than their non-medical, genetic nurse/genetic associate colleagues [14]. To what extent these reported differences in attitude reflect real differences in practice rather than awareness of the 'professionally correct' responses on questionnaires is uncertain. That the differences may be important is suggested by the pattern of results obtained in two earlier studies that have assessed the influence, in practice, of the background of the counselling professional (obstetrician or geneticist) on the continuation of a pregnancy in which the fetus has been identified as having a sex chromosome abnormality [15,16]; when counselling was provided by a geneticist the pregnancy was more likely to be continued.

Primary health care professionals are also less likely than geneticists to espouse non-directiveness, and studies using focus groups of primary care health professionals in the USA suggest that they have more doubts about the attainability of non-directiveness and are more willing than geneticists to provide definite guidance to their patients in relation to reproductive decisions [17]. It may therefore be that a professional adherence to non-directiveness (embedded in the ethos of genetic counselling but not of primary care) has an important influence on clinical practice even if, in reality, it is never fully attained.

FACTS, VALUES AND VALUE-NEUTRALITY

The goal of non-directiveness is closely related to the distinction between facts and values. Adherence to non-directiveness in genetic counselling may be seen as claiming that we professionals provide information and that the information is neutral; we are asserting that it is up to our clients how they make use of the information we provide. It is they who make all the decisions and therefore they carry all the responsibility. This is very comforting for us, but of course there is no such clear distinction between facts and values, between providing information and making decisions [18]. It is we professionals who decide what is to count as a fact, which facts to present, and how to present them; these decisions (perhaps only ever made implicitly or by default) impose a professional frame within which our clients are confronted with the restricted range of options from which we expect them to choose [19–21]. The influence of the framing of a decision upon the decision made is very clear within the context of genetics, as discussed in Chapter 6 on carrier screening [22].

In addition to the influence that we professionals exert upon our clients as individuals, there are broader professional factors to be considered. As health professionals, we may consider it to be in our interests to promote the growth of our specialty. This means that clinical geneticists may seek to increase the volume of genetic testing, and to include the adoption of new tests as technological developments permit (and perhaps before they have been properly evaluated). There may be enthusiasm for new tests simply resulting from an enthusiasm for the technology of testing; a 'can do – will do' mentality (see Chapter 6 on carrier screening). If we have close relationships with the biotechnology industry then we may even have a personal financial stake in the promotion of such testing. Furthermore, value-neutrality may be an attractive position for geneticists to adopt because it simplifies the decision-making processes about the introduction of a new genetic test. Such decisions could be left to clients, or to market forces, without any need for professionals to arrive at a consensus decision as to what testing should be made available. It would even be possible for simplistic, monetary cost–benefit analyses of genetic testing programmes to be justified if we adopt a value-neutral mentality in which nothing counts but 'the bottom line'; this could arise by default if a 'free' market is allowed to determine the shape of genetics services.

INFORMATION, COUNSELLING AND PRENATAL DIAGNOSIS

The information that would be relevant to any woman considering a prenatal genetic test encompasses social and emotional aspects that are often not discussed with our clients. Even the literature provided about conditions such as Down syndrome and cystic fibrosis, for which prenatal diagnosis and genetic screening are widely available, is significantly shaped by the prenatal context; it presents a substantially less positive image of the conditions than is portrayed in literature prepared for use by the parents of affected children (the 'twice-told tales' of Lipmann and Wilfond [23]). A decision *not* to address these broader social and emotional issues with our clients is to remove these opportunities for reflection from those who might find them helpful. To promote informed decisions, counsellors may feel it is appropriate to give information that has not been sought or to raise issues that our clients had not previously considered [24]. Giving priority to non-directiveness, in the negative sense of not influencing our clients, would then run counter to respect for informed decision-making. The context of genetic counselling is, perhaps sadly, too complex to permit adherence to any simple slogan as an adequate guide to our professional conduct [24].

It has been suggested that it is appropriate to draw to parents' attention the possible long-term consequences for their child if they decide to continue with a pregnancy in which the fetus has been found to be affected by a genetic condition such as Down syndrome [25]. In many contexts, this recommendation would clearly be directive [26]. Interview studies with genetic counsellors have shown that they are likely to challenge a woman who is carrying a Down syndrome fetus and seems unlikely to terminate the pregnancy by pointing out how severely affected the child may be; in contrast, they are much less likely to confront a woman who seems likely to opt for termination with optimistic statements about how healthy some affected children are and what good progress they make with appropriate programmes of developmental stimulation [27]. The idea of a 'balanced' presentation, as with party political comment on television current affairs programmes, is unlikely ever to be attained, but the present discrepancies between the information provided in different contexts is clearly unhelpful.

It may well be that the midwife or obstetrician and the pregnant woman offered prenatal screening tests collude by proceeding with the tests without an adequate discussion of the possible significance of the test results; a full discussion could be too emotionally demanding for either of them to want to face it [28,29]. As discussed earlier in this book, in Chapter 9 on prenatal screening, some antenatal clinics may offer prenatal genetic screening in such a routinized fashion that only rudimentary information is provided about the testing and only cursory consent is obtained.

The anguish, guilt, depression and regrets that can follow terminations of pregnancy for genetic disease or congenital anomalies are often not discussed in advance of screening or even in advance of a termination of pregnancy on such grounds [30,31]. The important possibility of prenatal diagnosis giving a confusing result (e.g. mosaicism) or an incidental finding of a relatively minor problem (e.g. Turner's syndrome or another sex chromosome anomaly), are also often not raised in advance of the prenatal test, and yet these possibilities should certainly be considered before an invasive prenatal diagnostic procedure is performed. The routine way in which these items of information are so often omitted from the professional portrayal of prenatal screening demonstrates the bias present in the clinical process. It is also most unlikely that the possibility of *actively welcoming* a child with Down syndrome is very often presented in antenatal clinics as a real ('reasonable') option, and if a woman makes that decision then personal values attributed to some unusual cultural or religious factors are likely to be invoked to account for her decision [27,32]. Values and 'culture' are not invoked to explain a decision to terminate an affected pregnancy. Providing information that puts Down syndrome in a positive light, as a way of balancing the usually negative portrayal of the condition, could cause distress to some women or couples; but to omit such information is effectively to nudge parents towards 'choosing' screening and towards terminating an affected pregnancy.

SOCIAL ISSUES AROUND GENETIC COUNSELLING

At the broader, societal level of analysis, we should also be addressing factors that will impact upon the decision-making process of our clients. For example, can we really regard a decision-making process as being non-directive if parents' fears of inadequate social or medical support for an affected child lead them to seek prenatal diagnosis? How can we describe a clinical process as being non-directive if the overall social context is clearly coercive? It cannot be possible to divorce the counselling from the social context. We can scarcely claim to be practising in a non-directive fashion if we fail to draw public attention to such gaps in social care provision.

Society and health professionals jointly decide to devote resources to prenatal genetic screening when alternative approaches to improving outcomes for mothers and infants are neglected, even though social and nutritional approaches to improving outcomes in high-risk pregnancies may be available and could be simpler and cheaper [33,34]. If we choose genetic approaches as opposed to environmental measures to improve the health of our infants, and if we fail to warn our clients about the potential adverse consequences of participation in the prenatal diagnostic process, then are we not behaving collectively in a directive manner? If 'value-neutrality' allows us to tolerate such implicit directiveness without demur then it indicates that we are functioning without values – amorally – rather than just impartially.

In addition to the effects of prenatal genetic screening on individual women who become caught up in such programmes, and who may suffer great anxiety as a result, there are also broader consequences of such screening programmes for all of us in society and for pregnant women in particular. As Rothman has described so well, the whole experience of pregnancy has been changed by the development of prenatal genetic screening so that it is now experienced as a 'tentative' state until the fetus has passed the various 'quality control' steps [35].

Another group who can be affected by prenatal genetic screening programmes are those who have the conditions that are to be 'prevented' by the screening. Many individuals with some learning disability are perfectly capable of understanding that prenatal screening programmes are being developed to prevent the births of other individuals like themselves in the future. This is bound to damage the already fragile self-esteem of many such individuals, and can cause great distress.

Is it reasonable to consider the effect of prenatal genetic screening programmes on those members of society who are affected by those conditions for which screening is made available? Remember that prenatal sex selection [by the termination of pregnancies in which the fetus is of the 'wrong', socially undesirable (usually female) sex] is regarded as ethically unacceptable in many countries, including Britain. This clearly reflects the relatively high status of women in modern British society. Then contrast this with the situations that arise in the context of genetic screening and counselling, where terminations of pregnancy are available for fetuses even at low risk of a serious disorder, at some risk of mild–moderate learning problems, or who are definitely affected by a minor physical anomaly (such as a cleft lip).

Why can terminations of pregnancy be made available for such problems? By comparison with fetal sex selection, we can only conclude that the social status of those with even minor disabilities is regarded as too insignificant for the offence it causes to be taken into account. The lobby of enthusiasts promoting prenatal screening for Down syndrome certainly seems to have made no effort to take any account of the views of individuals with Down syndrome, or of those for whom a child with Down syndrome would not be regarded as a problem [36,37]. When stated in such stark terms, is it not shameful that the views of those for whom (ostensibly) the screening programme is being provided are not taken into account? Does it not lay us open to the charge that the purpose of such screening is not to avoid suffering or to facilitate informed choice but rather something quite else: perhaps to reduce health care expenditure or to promote a particular technology?

The reluctance of many professionals to discuss this issue can be understood; if we did decide that prenatal testing and terminations of pregnancy were indeed inappropriate for certain conditions, because affected individuals felt strongly that this approach dehumanized them and contributed to their exclusion from full participation in society, then how would we respond to requests from couples who wished to have prenatal diagnosis because of their risk of an affected child? And how would we take account of the views of those individuals affected by genetic conditions who have suffered badly from medical or social problems and who adopt the view that their lives have not been worthwhile and that termination of a pregnancy would be preferable to giving birth to another affected child? How could we draw up a list of conditions for which testing, screening or termination would be regarded as inappropriate?

It is difficult to imagine how a social consensus would be arrived at; it would certainly require a political process that involved society as a whole and not just health professionals. The difficulty of imagining how such a public debate could be resolved has served to frighten us from even beginning to raise the issues; when they have been raised, some of my fellow geneticists have reacted vigorously, as if a taboo had been violated. Perhaps this is the case; sex, death and race are no longer the taboo areas that they once were, and perhaps disability has succeeded them as the most socially sensitive issue.

PAUSE FOR REFLECTION

It can be seen that genetic counsellors espouse non-directiveness as a goal, but that there are numerous problems with the applications of this concept. We can see that the process of genetic counselling, especially in population screening programmes, can be heavily directive, both as a result of the institutional context and because individual counsellors are uncomfortable with providing even-handed information about genetic conditions for which termination of pregnancy is current standard practice. Counsellors may find it helpful for a number of reasons to lay claim to non-directiveness; it is functional for them, especially when there may be a conflict of values between counsellor and client [27,37], but that does not mean that genetic counselling services actually function in a non-directive manner.

Furthermore, by espousing non-directiveness so vehemently, we genetic counsellors may disqualify ourselves from participating in the much-needed public debates about what types of genetic testing should be made available in our various countries, and through what channels genetic tests should be provided [38,39]. As the professional group most heavily involved in genetic testing and most experienced in the outcomes of genetic tests, it is clearly necessary for us to contribute to discussions about what tests are appropriate and how they should be provided. Our experience and judgement will be important in helping society decide about the appropriate regulation of commercially driven testing programmes, and about prenatal or predictive testing for late-onset disorders and prenatal testing for minor conditions.

ASSESSING DIRECTIVENESS AND NON-DIRECTIVENESS

Our discussion of non-directiveness has explored this term from two perspectives. We have looked at why genetic counsellors may seek to describe their activities as being non-directive, and we have considered wider social influences that may constrain or influence individuals' decisions. We have not examined what happens in genetic counselling to see whether, or to what extent, it is non-directive in practice. There have been very few attempts to study the process of genetic counselling for objective evidence of directiveness or non-directiveness.

Kessler commented on a transcript of one fairly unremarkable genetic counselling interview; he pointed out that the counsellor appeared to respect non-directiveness but that his overall approach was probably encouraging his client to accept prenatal diagnosis [6]. More than 10 years later, Kessler returned to the issue of non-directiveness [40]. He made the point that all genetic counselling is aimed at altering the client's behaviour; directive counselling aims to lead the client to make a particular decision, whereas non-directive counselling aims to help the client to make a decision 'well', not to prejudge the outcome of the decision. He repeated his call for further research into what actually happens within the genetic counselling encounter; very little such research has in fact been undertaken. As discussed in Chapter 12 on outcomes and process, it is clear that most measures of outcomes from genetic counselling are unsatisfactory but that analyses of the process of genetic counselling have been too few and too rudimentary.

The research group of Marteau and her colleagues has recently set out to study the relationship between the process of genetic counselling and its outcomes, and in particular to examine the issue of non-directiveness. Despite studying a range of client and counsellor characteristics and their expectations of a genetic counselling session across a series of 131 genetic counselling sessions, these variables showed little effect on the outcome measures recorded following each consultation, such as satisfaction and anxiety. Furthermore, none of the measures of process used to describe the process of counselling predicted the outcomes for the client [41]. Given my argument in Chapter 12, this is no surprise. Indeed, it would have been a great surprise if the chosen measures of outcome had revealed any significant effect of the process of counselling. While it could be that genetic counselling is a mechanical process with little opportunity for variation in quality or effectiveness between counsellors or counselling sessions, so that the main predictor of outcome is input, there is another possibility; that the methods of analysis adopted in this study were too crude to identify the important features of the consultations. Those attached to measures of outcome, and convinced of their utility, may favour the former interpretation of this study; practitioners like myself will probably favour the latter.

As a part of this study, the recording of clients' expectations was conducted by presenting them with a predetermined list of possible expectations which genetic counsellors had previously generated as valid aims of their service. It was therefore not possible to address the issue of counsellors finding it necessary to modify rather than meet their clients' inappropriate expectations [42]. This is a crucial aspect of genetic counselling that needs to be examined systematically in the future; both the process of the counsellor coming to a decision that the client's expectations are inappropriate and the means by which the counsellor attempts to modify these expectations deserve detailed investigation (see Chapter 12).

A further objective of the analysis of 131 genetic counselling sessions was to assess the directiveness of the counsellor from three different perspectives (as perceived by the client, as rated by a researcher and as reported by the counsellor) and to relate directiveness to process variables and to outcomes of the genetic counselling [43]. Directiveness was therefore assessed on a scale, rather than as an all-or-nothing phenomenon. Directiveness was not shown to influence the outcomes of counselling in terms of the meeting of expectations, satisfaction, anxiety or concern. Several factors (patterns of speech interaction during the session, the socio-economic status of the client, counselling skills training of the counsellor and the counsellor's judgement of the client's level of concern) did influence the directiveness of the counsellor as rated by the researcher. Directive statements were categorized by the researcher as 'advice', 'evaluation' or 'reinforcement', and these were influenced separately by some of the relevant factors. The only relation identified between the different measures of directiveness was an association between counsellor-reported directiveness and researcher-rated reinforcement.

The validity of the conclusions of this study depend critically upon the authors' system of rating directiveness [43]. The very few examples provided of statements scored as being directive do raise some questions about the method of interpretation. Thus, the statement, "There's a very good chance that you will have another healthy baby", is categorized as directive (and evaluative) but it is not possible to come to an independent opinion about this when the sentence is decontextualized; the statement could be merely informative. Similarly, the statement, "It would be sensible if you spoke to Michael and

Carol about this", is classed as being directive (and advisory). There are many contexts in which such a statement would be helpful and thoroughly appropriate within genetic counselling, and not even the most ardent advocate of non-directiveness would want to forbid such remarks. Again, it need not be directive to utter "I understand; that's really very sensible", but this was also scored as directive (and reinforcing).

The operational definition of directiveness within this study, therefore, may be misleading. It may distract us from the real issues concerning the ways in which inappropriate influences are brought to bear on clients, and may make it more difficult for us to confront these inappropriate interventions in genetic counselling. While Michie, Marteau and colleagues are to be praised for tackling such a theme, we may not gain much enlightenment until a more sophisticated framework of analysis has been developed. As proposed above and in the previous chapter, this will need to focus on the process of genetic counselling and on the way in which clients' inappropriate expectations are modified by this process.

APPROPRIATE DIRECTIVENESS?

One area of genetic counselling in which a strong adherence to non-directiveness may be misplaced relates to predictive, presymptomatic testing and susceptibility testing, at least where there are medical interventions or life-style changes that may modify the course of the disease or the probability of developing health problems. Genetic testing offered in such a context may legitimately be recommended as likely to improve the health of the individual. This assumes, of course, that the test of genetic susceptibility has indeed been shown to be of benefit to those tested (see Chapter 7 on population screening for genetic susceptibility to disease). The offer of testing and the associated counselling should of course promote informed decision-making by the client, but it is then in effect a form of standard medical advice-giving and there would be no reason for the genetic counsellor to wear a mask of strict neutrality [44,45]. The most appropriate approach to this type of counselling may be an active partnership model [46] in which professional and client share control and responsibility. The characteristics of such a model of shared decision-making still need to be refined [47].

Within a counselling session, some counsellor behaviours may be perfectly appropriate even though they may be classed as directive from certain perspectives [43]. In relation to genetic testing issues, it may be very helpful to encourage a client to consider the likely consequences for themselves and others of the various possible decisions they might make (e.g. not to be tested, to be tested and have a favourable result, to be tested and have an unfavourable result). Such encouragement may be seen as being directive, leading the client to consider implications that he/she would not have otherwise confronted. An excessive attachment to non-directiveness could paralyse this counselling component of genetic counselling, in which the confrontation of clients with information or with possible future scenarios may be an essential element.

BEYOND NON-DIRECTIVENESS

Because the meanings of the words 'directiveness' and 'non-directiveness' can be confused or contentious, and because the issue of directiveness can apply to a range of contexts from large-scale social phenomena to the minutiae of interactions within a counselling session, we should perhaps avoid these words altogether. This has certainly been advocated by Wolff and Jung, who favour an approach to genetic counselling which starts from the experience and background of the clients [48]. This approach recognizes the dependence of both client and counsellor on their situation and on the information available to them. This approach values direct, concrete communication between client and counsellor, which may include the recognition and challenging of blocks to communication and of conflict within a family or within the counselling session. It recognizes that it may be appropriate for the counsellor to control some aspects of the counselling session, such as information-gathering and the definition of the range of scenarios to be considered as possible outcomes of any decision-making process.

Wolff and Jung's approach to genetic counselling draws upon Rogers' concept of client-centred therapy but moves beyond it to focus on the meaning and personal significance of the issues facing the client. Genetic counselling, then, will inevitably contain information as a core component, but incorporates elements of a psychotherapeutic process. Instead of trying not to influence clients (as if that were possible!) the counsellor tries to influence the client in an essentially therapeutic manner. This places emphasis upon a concern for the meaning of the genetic issues confronting a client for their concept of themselves and their life tasks. It demands continual and systematic reflection upon one's role as a genetic counsellor. It incorporates elements of non-directiveness, because the counsellor must be free of the need to contribute his/her own attitudes and value judgements, but it moves beyond simple non-directiveness.

DESCRIPTIONS OF GENETIC COUNSELLING

If we are to improve genetic counselling practice, we need to develop a richer description of the process of genetic counselling and of the impact of genetic disease and of genetic counselling on individuals and families. This will help us to understand more thoroughly the task in which we are engaged, and then to identify ways of improving our practice. These studies of genetic counselling and its social impact will not be of relevance solely to genetic counsellors, however, but will be of real importance in many areas. This is because the issues, with which we in clinical genetics have been confronted for some years, are becoming of much broader relevance as we gain the ability to identify the genetic factors contributing to so many common disorders. The problems that now arise in genetic counselling and its impact on families with uncommon genetic diseases will soon be commonplace in primary health care and general medical practice. Our ability to identify and resolve the problems we are now facing will provide an invaluable guide to how society should approach many of the areas that will produce real difficulties in the future. These issues form the core of this volume, and it is their relevance beyond genetic counselling itself that justifies the wide public and professional interest they have aroused.

To generate a richer description of genetic counselling and of communication about genetic issues within families, therefore, will be important outwith the narrow confines of the genetic counselling profession. There are two principal perspectives that will be of central importance in developing this richer description of the process and social impact of genetic counselling. First, there is the group of academic disciplines that examines the whole area of communication and language. This draws upon linguistics, sociology, literary theory and social psychology in order to describe and understand all types of discourse, conversation, text and communication. These studies hold out the promise of a rigorous, objective analysis of the process of communication between counsellor and client as well as between professionals [49] and between family members outside the clinic. Secondly, there are the practitioners of counselling and psychotherapy who are able to provide feedback to professionals on the processes of interaction within a clinic setting, and to guide the professional to a better understanding and response to these interactions for the sake of the clients and their families.

There have been few studies of genetic counselling from the perspective of sociolinguistics or other communication theories. The most sophisticated model of genetic counselling practice that I have yet seen has been developed by Jennifer Hartog from her observations and discourse analysis of 33 audio-taped counselling sessions [50]. Her understanding of the pattern of genetic counselling practice has been derived from empirical observation and analysis rather than from accounts of what professionals think is happening or ought to be happening. It is striking (see *Figure 13.1*) that the process of interaction between counsellor and client nearly ceases once the counsellor has defined the problem and gathered information from the client. From that point, it seems that the counsellor often passes information to the client, including risk information, and then closes the session by giving advice about what needs to be done next. The client seems to make relatively little contribution to the discussion once the information-gathering by the counsellor is complete. This may reflect the nature of the cases included in Hartog's series, or may reflect the practice in that specific clinic. It may be, however, that there is a more

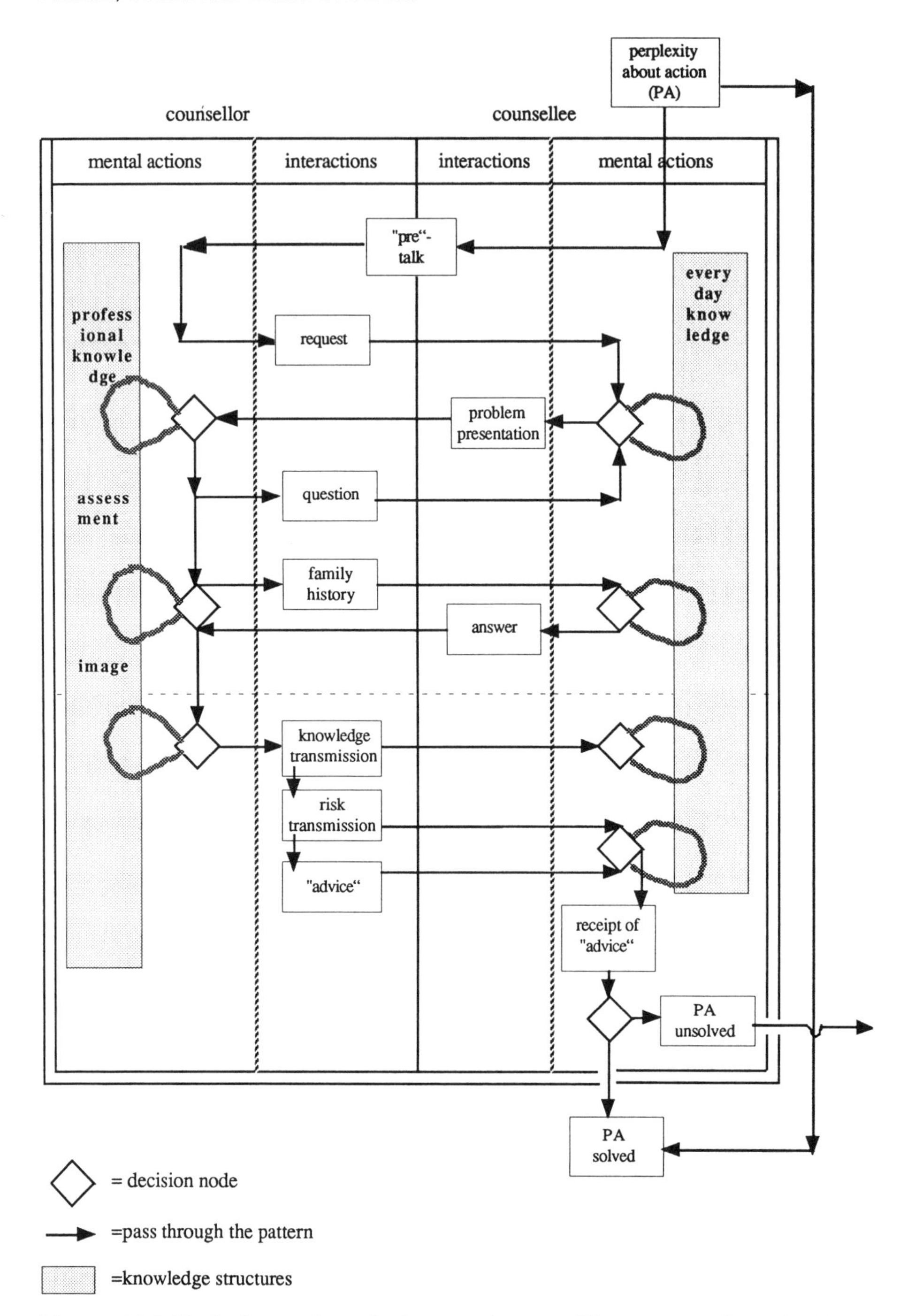

Figure 13.1. Typical pass through the genetic counselling pattern (Jennifer Hartog, Das genetische Beratungsgespräch. Institutionalisierte Kommunikation zwischen Experten und Nicht-Experten. Gunter Narr Verlag, Tübingen, 1996).

general issue – there may be a mismatch between our (professional) representation of what we do and what actually happens in our clinics.

Another approach to the analysis of genetic counselling process is to identify recurrent patterns of dialogue, and how these patterns are used by counsellors to structure the counselling sessions. Although this has not been carried out very much in relation to genetic counselling, it has been developed in related areas such as the giving of potentially unwelcome information to parents about a serious problem identified in their child. Maynard uses conversation analysis to describe the perspective-display series in which a clinician asks the child's parent(s) for his/her views of the child's problems, the parent responds to that query and the clinician then delivers the relevant diagnostic information in a form that is (hopefully) tailored to the parent's perspective [51]. A systematic description of such tools in the repertoire of the clinician or counsellor enables the differences between individual performances to be highlighted, and the explanations for such case-specific features to be sought in relation to counsellor–client interactions or other factors. Such analyses converge with those that derive from the various schools of psychotherapy and counselling, and studies of genetic counselling that draw upon both approaches will be very productive.

The first requirement for developing our powers of description of the genetic counselling process entails developing a common vocabulary to describe the functions (effects rather than intended purposes) of the various statements made by the professional in a genetic counselling session. One possibility would be to refine the operational definitions of the subcategories of directiveness developed by Michie and colleagues [43], but this framework may be too restricted to the issue of directiveness, and it may be more constructive to move beyond the directive/non-directive dichotomy. The most suitable framework for the description of counsellor statements and interventions that I have worked with is the 'six category intervention analysis' approach developed by John Heron [52]. This categorizes all counsellor interventions into three authoritative and three facilitative categories. The authoritative categories are the prescriptive, the informative and the confronting. The facilitative categories are the cathartic, the catalytic and the supportive. Heron's development of this framework provides an accessible vocabulary that permits practitioners to discuss their own practice and that permits feedback from observers and colleagues, and which is also suitable for research-oriented studies of lay–expert encounters; this framework may help the worlds of clinical practice and academic discourse analysis to interact. Let us hope that this can be fruitful.

CONCLUSIONS

As Shiloh says, genetic counsellors (should be) " ... helping clients reach a decision wisely, rather than reach a wise decision" [53]. This requires us to focus on the theoretical issues and counselling techniques that relate to the making of decisions. It also requires us to develop our understanding of the social context within which individuals and families live their lives and within which they make their decisions. By broadening our consideration of the process of genetic counselling, we can move beyond the focus on the directive/non-directive dichotomy. This has been useful but is now too restricting.

It is possible to approach the study of genetic counselling with the intention of identifying those aspects of the within-clinic interaction that predict outcome [54], but I would argue that this approach is too narrow for three reasons. First, it could only ever be possible to 'predict' outcomes that would be too crude to satisfy either client or professional that they reflected the value of their consultation, and the 'prediction' could only ever amount to a statistical observation rather than a true prediction in an individual case. That may be enough to plan public health resource allocations, but is inadequate at the level of individual clients. Second, this approach ignores the social context within which the genetic counselling process is necessarily embedded. Third, and more philosophically, within such an approach is the implicit image of man as a mechanical construct in which consciousness is a mere epiphenomenon and an individual's behaviour is determined by the rigid laws of the physical universe [55]. This is an inadequate and shallow understanding of our biological and social nature.

Combining the perspectives of a practical-based psychotherapy with a more abstract discourse analysis will allow us to dissect and comprehend the processes of genetic counselling, although it will not lead to predictions of patient outcomes. The focus of Wolff and Jung on the personal experience of the client allows the human meaning of the client's experience to remain as a central concern of the counselling, as in the use of narrative analysis to identify those individuals and families likely to experience the greatest problems in adjusting to their situation and hence require the most intensive support [56]. Such an approach will also prevent us from making unwarranted assumptions about the value and significance of numerical risk figures when these are given to clients. Of course professionals place a great emphasis upon the accuracy of numerical information given to clients, but the effort expended in generating this precision will not always be justified. The significance of measures of susceptibility to heart disease, for example, may be distorted while their significance is mediated to patients through a health professional, and precise estimates (!) of risk may be harmful in leading to an unhelpful preoccupation with illness in individuals who are essentially healthy [57].

Understanding the processes of communication between counsellor and client in genetic counselling is obviously important to improving our practice, but if we are to apply this understanding to society at large (which will become ever more important as the contribution of genetics to health care continues to expand) then we must attend on the social context within which the process of genetic counselling occurs. We must also attend to:

(i) the communication of information about genetic conditions and genetic risks between family members,
(ii) the processes of communication about genetic counselling and testing between professional colleagues (between genetic counsellors, clinicians and laboratory scientists),
(iii) the social and emotional impact of developments in genetics, especially programmes of genetic screening and testing, on those with genetic conditions including the condition of being 'at-risk',
(iv) thc public representation in schools and the media of diseases and conditions as being 'genetic', and the influence of these representations on the willingness of society to provide social and medical support for affected individuals.

Careful research studies of current genetic counselling and screening services can help to improve professional practice to the benefit of our clients at the same time as the research illuminates important social processes and guides future public policy. A careful analysis of the processes and problems that have already become manifest within genetic counselling and screening will allow us to avoid many of the potential future problems that could arise from the over-hasty and ill-considered misapplication of human genetics to health care and to social problems.

REFERENCES

1. Connor M, Ferguson-Smith M (1997) *Essential Medical Genetics*, 5th Edn, p. 103. Blackwell, Oxford.
2. Fraser FC (1974) Genetic counselling. *Am. J. Hum. Genet.* **26**: 636–659.
3. Clarke A (1990) Genetics, ethics and audit. *Lancet*, **335**: 1145–1147.
4. Harper PS (1992) Genetics and public health. *Br. Med. J.* **304**: 721.
5. Clarke A (1991) Is non-directive genetic counselling possible? *Lancet*, **338**: 998–1001.
6. Kessler S (1981) Psychological aspects of genetic counselling: analysis of a transcript. *Am. J. Med. Genet.* **8**: 137–153.
7. Marteau TM, Drake H (1995) Attributions for disability: the influence of genetic screening. *Soc. Sci. Med.* **40**: 1127–1132.
8. Burke BM (1992) Genetic counsellor attitudes towards fetal sex identification and selective abortion. *Soc. Sci. Med.* **34**: 1263–1269.
9. Bosk CL (1992) *All God's Mistakes. Genetic Counseling in a Pediatric Hospital.* University of Chicago Press, Chicago and London.
10. Wertz DC, Fletcher JC (1988) Attitudes of genetic counselors: a multinational survey. *Am. J. Hum. Genet.* **42**: 592–600.
11. Wertz DC, Fletcher JC (1989) *Ethics and Human Genetics: a Cross-cultural Perspective*. Springer, Heidelberg.

12. Wertz DC (1992) Ethical and legal implications of the new genetics: issues for discussion. *Soc. Sci. Med.* **35**: 495–505.
13. Marteau TM, Drake H, Reid M, Feijoo M, Soares M, Nippert I, Nippert P, Bobrow M (1994) Counselling following diagnosis of fetal abnormality: a comparison between German, Portuguese and UK geneticists. *Eur. J. Hum. Genet.* **2**: 96–102.
14. Marteau T, Drake H, Bobrow M (1994) Counselling following diagnosis of a fetal abnormality: the differing approaches of obstetricians, clinical geneticists and genetic nurses. *J. Med. Genet.* **31**: 864–867.
15. Holmes-Siedle M, Ryyanen M and Lindenbaum RH (1987) Parental decisions regarding termination of pregnancy following prenatal detection of sex chromosome abnormality. *Prenat. Diag.* **7**: 239–244.
16. Robinson A, Bender BG, Linden MG (1989) Decisions following the intrauterine diagnosis of sex chromosome aneuploidy. *Am. J. Med. Genet.* **34**: 552–554.
17. Geller G, Holtzman NA (1995) A qualitative assessment of primary care physicians' perceptions about the ethical; and social implications of offering genetic testing. *Qual. Hlth Res.* **5**: 97–116.
18. Gervais KG (1993) Objectivity, calue neutrality and Nondirectiveness in Genetic Counseling. In: *Prescribing Our Future: Ethical Challenges in Genetic Counseling*, (eds D.M. Bartels, B.S. LeRoy, A.L. Caplan), pp. 119–130. Aldine de Gruyter; New York.
19. Tversky A, Kahneman D (1981) The framing of decisions and the psychology of choice. *Science,* **211**: 453–458.
20. Shiloh S, Sagi M (1989) Effect of framing on the perception of genetic recurrence risks. *Am. J. Med. Genet.* **33**: 130–135.
21. Marteau TM (1989) Framing of information: its influence upon decisions of doctors and patients. *Br. J. Soc. Psychol.* **28**: 89–94.
22. Bekker H, Modell M, Denniss G, Silver A, Mathew C, Bobrow M, Marteau T (1993) Uptake of cystic fibrosis testing in primary care: supply push or demand pull? *Br. Med. J.* **306**: 1584–1586.
23. Lippman A, Wilfond BS (1992) Twice-told tales: stories about genetic disorders. *Am. J. Hum. Genet.* **51**: 936–937.
24. Singer GHS (1996) Clarifying the duties and goals of genetic counselors: implications for nondirectiveness. In: *Morality and the New Genetics. A Guide for Students and Health Care Providers* (eds B. Gert, E.M. Berger, G.F. Cahill, K.D. Clouser, C.M. Culver, J.B. Moeschler, G.H.S. Singer), pp. 125–145. Jones and Bartlett Publishers, Sudbury, MA.
25. Nuffield Council on Bioethics (1993) *Genetic Screening – Ethical Issues*. Nuffield Council on Bioethics, London.
26. Clarke A (1994) Genetic screening: a response to Nuffield. *Bull. Med. Ethics,* **940** (April): 13–21.
27. Brunger F, Lippman A (1995) resistance and adherence to the norms of genetic counseling. *J. Genet. Counsel.* **4**: 151–167.
28. Press NA, Browner CH (1994) Collective silences, collective fictions: how prenatal diagnostic testing became part of routine prenatal care. In: *Women and Prenatal Testing: Facing the Challenges of Genetic Technology* (eds K.H. Rothenberg, E.J. Thomson), pp. 201–218. Ohio State University Press, Columbus.
29. Browner CH, Press NA (1995) The normalization of prenatal diagnostic screening. In: *Conceiving the New World Order: The Global Politics of*

Reproduction (eds F.D. Ginsburg, R. Rapp), pp. 307–322. University of California Press, Berkeley, CA.
30. White-van Mourik MCA, Connor JM, Ferguson-Smith MA (1992) The psychosocial sequelae of a second-trimester termination of pregnancy for fetal abnormality. *Prenat. Diagn.* **12**: 189–204.
31. White-van Mourik M (1994) Termination of a second-trimester pregnancy for fetal abnormality. Psychosocial aspects. In: *Genetic Counselling: Practice and Principles* (ed. A. Clarke), pp. 113–132. Routledge, London.
32. Clarke A (1997) *Culture, Kinship and Genes* (A. Clarke, E.P. Parsons, eds). Macmillan, London (in press).
33. Lippman A. (1991) Prenatal genetic testing and screening: constructing needs and reinforcing inequities. *Am. J. Law. Med.* **17**: 15–50. Reprinted (1994) in *Genetic Counselling: Practice and Principles* (ed. A. Clarke), pp. 142–186. Routledge, London.
34. Lippman A (1992) Mother matters: a fresh look at prenatal genetic testing. *Issues Reprod. Genet. Eng.* **5**: 141–154.
35. Rothman BK (1988) *The Tentative Pregnancy: Prenatal Diagnosis and the Future of Motherhood.* Pandopra Press, London.
36. Goodey CF (ed.) (1991) *Living in the Real World. Families Speak About Down's Syndrome.* The Twenty-One Press, London.
37. Rapp R (1988) Chromosomes and communication: the discourse of genetic counselling. *Med. Anthr. Quart. (New Series)* **2**: 143–157.
38. Burke BM, Kolker A (1994) Directiveness in prenatal genetic counseling. *Women Hlth,* **22**: 31–53.
39. Caplan AL (1993) Neutrality is not morality: the ethics of genetic counseling. In: *Prescribing Our Future: Ethical Challenges in Genetic Counseling* (eds D.M. Bartels, B.S. LeRoy, A.L. Caplan), pp. 149–165. Aldine de Gruyter; New York.
40. Kessler S (1989) Psychological aspects of genetic counseling. VII Thoughts on directiveness. *J. Genet. Counsel.* **1**: 9–17.
41. Michie S, Axworthy D, Weinman J, Marteau T (1996) Genetic counseling: predicting patient outcomes. *Psychol. Hlth,* **11**: 797–809.
42. Michie S, Marteau TM, Bobrow M (1997) Genetic counselling: the psychological impact of meeting patients' expectations. *J. Med. Genet.* **34**: 237–241.
43. Michie S, Bron F, Bobrow M, Marteau TM (1997) Nondirectiveness in genetic counselling: an empirical study. *Am. J. Hum. Genet.* **60**: 40–47.
44. Motulsky AG (1994) Predictive genetic diagnosis (Invited Editorial). *Am. J. Hum. Genet.* **55**: 603–605.
45. Bernhardt BA (1997) Empirical evidence that genetic counselling is directive: where do we go from here? (Invited Editorial). *Am. J. Hum. Genet.* **60**: 17–20.
46. Roter D (1987) An exploration of health education's responsibility for a partnership model of client–provider relations. *Patient Educ. Counsel.* **9**: 25–31.
47. Charles C, Gafni A, Whelan T (1997) Shared decision-making in the medical encounter: what does it mean? (or it takes at least two to tango). *Soc. Sci. Med.* **44**: 681–692.
48. Wolff G, Jung C (1995) Nondirectiveness and genetic counselling. *J. Genet. Counsel.* **4**: 3–25.
49. Atkinson P (1995) *Medical Talk and Medical Work.* Sage, London.
50. Hartog J (1996) *Das Genetische Beratungsgespräch. Institutionalisierte Kommunikation zwischen Experten und Nicht-Experten.* Gunter Narr Verlag, Tübingen.

51. Maynard DW (1991) The perspective-display series. In: *Talk and Social Structure. Studies in Ethnomethodology and Conversation Analysis*. (eds D. Boden, D.H. Zimmerman), pp. 164–192. Polity Press, Cambridge.
52. Heron J (1989) *Six Category Intervention Analysis*, 3rd Edn. Human Potential Resource Group, University of Surrey, Guildford.
53. Shiloh S (1996) Decision making in the context of genetic risk. In: *The Troubled Helix. Social and Psychological Implications of the New Human Genetics* (eds T. Marteau, M. Richards), pp. 82–103. Cambridge University Press, Cambridge.
54. Michie S, Marteau T (1996) Genetic counselling: some issues of theory and practice. In: *The Troubled Helix. Social and Psychological Implications of the New Human Genetics* (eds T. Marteau, M. Richards), pp. 104–122. Cambridge University Press, Cambridge.
55. Shotter J (1975) *Images of Man in Psychological Research* (Vol. F7, Essential Psychology series). Methuen, London.
56. Brock SC (1995) Narrative and medical genetics: on ethics and therapeutics. *Qualit. Hlth Res.* **5**: 150–168.
57. Adelsward V, Sachs L (1996) The meaning of 6.8: numeracy and normality in health information talks. *Soc. Sci. Med.* **43**: 1179–1187.

Chapter

14

GENETIC RESEARCH AND 'IQ'

P.S. Harper

INTRODUCTION

Intelligence is undoubtedly among the most important and distinctive human attributes, and argument over the nature and extent of its biological basis is nothing new. Scientific debate has frequently become entangled with racial issues and with political views, while the basis of measuring intelligence has been disputed, with accusations of fraud in relation to one well-known past expert.

Not being a psychologist nor having myself any other claim to expertise in this field, readers may wonder why this topic is included in a book on social and ethical issues arising in clinical genetics practice; the answer simply is that the issue arose, quite unexpectedly, in our own Cardiff department, and that we were forced to confront it, even though we had not previously given the matter serious thought.

The very unexpectedness of this situation is perhaps a valid reason for including it; clinicians are inevitably going to have to confront new and unforeseen issues arising in practice, and the way in which we handled this particular one may be instructive to others. This short chapter links closely with the following one; though written independently of it, the concerns are similar.

The Cardiff Institute of Medical Genetics has had a long-standing interest in neurogenetics, with a fruitful and growing collaboration in psychiatric genetics resulting from the specific research interests of our Department of Psychological Medicine. This joint research ranges from the genetic basis of schizophrenia and manic depressive illness, to Huntington's disease; the molecular genetic and gene mapping expertise is increasingly focused on identifying specific genes in common, non-Mendelian disorders such as the major psychoses.

Parallel work originating in the USA has developed on identifying specific genes involved in the normal processes of behaviour and cognitive function. This 'quantitative trait loci' (QTL) approach, combining appropriate mathematical analyses with molecular techniques, offers particular opportunities in this difficult field.

In 1993, The Institute of Medical Genetics found itself indirectly involved in a collaborative study initially co-ordinated from the USA, on the genetic basis of cognitive ability, a term approximately equivalent to what most people would term 'IQ' (intelligence quotient). This link arose from two of our Psychological Medicine colleagues involved in this project having joint appointments with Genetics, and from the laboratory base initially being sited in our Institute. It was only when the Institute of Medical Genetics was credited in the author list of a preliminary paper [1] (not showing any definitive results), that most staff became aware of the research, and a sense of unease developed that we had become involved in a sensitive area without proper discussion. The fact that the project was primarily based in the USA, that the subjects involved (a series of normal school children) were American, and that our laboratory was simply involved in analysing some of the genetic markers, explained how the situation had arisen, but did not lessen the unease.

The upshot was to hold a debate involving all scientific and clinical staff, as well as colleagues from both departments involved in the project. The result was most encouraging: the difficulties of work in the field were recognized, the potential dangers raised and the whole subject was discussed in a constructive and friendly manner. It was clear, though, that genuinely different views existed, even though everyone had gained by considering the different viewpoints. At the end of the day, a decision had to be made, and this was that Medical Genetics, as an academic department, should dissociate itself from the work, at least until some basic issues had been resolved. The following letter was written to the journal that had published the original paper. I am not aware that it was followed by any published replies or comments.

DNA MARKERS ASSOCIATED WITH HIGH VS. LOW IQ: ETHICAL CONSIDERATIONS[a]

The recent publication of the article on DNA markers and high and low IQ, by Plomin *et al.* (1994), has resulted in considerable debate in our Department of Medical Genetics in Cardiff and prompts the writing of this letter. The close collaborative work of the two departments in a number of other areas, together with indirect involvement from the fact that some of the DNA analyses in the published paper were undertaken in a laboratory located within our institute, has raised concerns, which have recently taken shape in the form of specific discussion and debate on the ethical issues involved. The constructive nature of this debate and the hope that it may help others in this difficult field are the main reason for writing this letter.

My concerns fall into four broad areas:

---

[a]Reproduced from *Behavior Genetics* (1995) by kind permission.

(i) whether the research itself on genetic markers and IQ can be considered ethical.
(ii) Whether misuse of any results is possible or likely in the future.
(iii) Whether the research might have adverse implications for the more medically orientated research and service activities of our two clinically based departments.
(iv) The potential for misinterpretation of any results, and of the research itself, by the media and the public.

I should like to consider each of these points in turn.

(i) I do not myself consider this research necessarily to be unethical, but there are undoubtedly major and controversial issues involved, and I think that it is unethical to undertake the work without giving these the fullest consideration. Any results involving a specific gene or marker relating to IQ are likely to have a profound significance and to alter people's perception of the underlying basis of intelligence in a way quite different from that resulting from more general studies of the genetic component of intelligence. I think that it is essential that there should be widespread professional and public consultation over this work.

A point of particular concern, with important general implications, is that the local ethical review committee of the Health Authority, to whom the study has been submitted, has taken the view that the work is not of direct concern to it, since the subjects involved live in the United States and are not patients. I would submit that such committees have a duty to consider the general ethical aspects of research, as well as the direct implications to patients that are the primary remit of such committees. No central body exists in Britain whose function is to consider such general issues.

(ii) The practical implications and future applications of this research are uncertain. The authors point out clearly that their work is unlikely to form the basis of future tests of prediction, because of the probable small effect of individual loci. However, for an attribute regarded as so important by both society and many families, I do not think that such a conclusion can be taken for granted; there could well be pressure to apply (or misapply) any marker or series of markers found to be relevant, even without an adequate scientific basis for doing so. The powerful impact of identifying specific genes involved in the normal variation of intelligence, even if their individual effect is small, as opposed to a more general involvement of unidentified genes, should not be underestimated.

(iii) Both the Department of Medical Genetics and the Department of Psychological Medicine at our University are clinically based and orientated. Most of our research is on genetic and psychiatric disorders, while both departments also have a major service role in their respective fields. I strongly support the involvement of clinical departments in fundamental research – we have ourselves been involved in the isolation of a series of major disease genes – but it is essential to consider the possible effects of involvement in a potentially controversial basic area on the more clinical aspects of our work. In particular, the juxtaposition of such research with more clinical activities might produce an impact that would not occur if the basic research were solely to involve workers without any clinical involvement.

(iv) However objectively and cautiously the results of this research are presented and published, the finding of a clearly positive result in relation to a specific genetic locus and intelligence will result in intense media publicity. Given the tendency to sensationalize and trivialize that is so widespread, the way in which any such result reaches the general public will probably be beyond the control of the investigators. In my opinion, there is a distinct possibility of not only the work itself, but the more general areas of genetic research and genetic testing receiving adverse publicity, something that would harm those trying to ensure the responsible application of these developments, and that might increase the fears of the public about the work of scientists and clinicians in this field, especially given the history of eugenic abuse in psychiatric genetics.

The recent debate and discussion between members of our two departments on this subject have been extremely constructive, though it would have been even more helpful if it had occurred before the work was begun. As a result of the discussions and of the continuing concerns with the Department of Medical Genetics, I have decided, in consultation with colleagues, that this department cannot be involved in or associated with the work until the ethical issues are clarified and hopefully resolved. This should be seen as applying not only to the paper published, but also to other ongoing aspects of the research.

My colleagues and I hope that the writing of this letter, in consultation with my colleagues in the Department of Psychological Medicine, will form the basis of informed and constructive discussion within the research community. The fact that locally we have now been able to have a critical yet constructive debate on this difficult issue, while continuing as close collaborators in other research fields, is heartening and will, I hope, encourage others involved in equally difficult fields to engage in open dialogue regarding any controversial aspects of their work.

Readers may be interested to know how the subject has developed since the letter was published in 1995. In essence, little has changed. Collaboration in psychiatric genetics between our two departments continues to flourish and grow, the original project has been maintained (but not involving Medical Genetics), with its focus now in London since the recent move there from the USA of the principal investigator. Our local ethical committee eventually considered the topic (with considerable reluctance!) and had no objections to it, while the UK Medical Research Council (equally reluctant to address the issue!) has now belatedly recognized that this field of work could cause public concern and should be discussed widely and openly [2,3]. No clear positive results have emerged yet from the research (Spring 1997), while media attention has temporarily been diverted towards 'delinquency genes', the genetic basis of homosexuality, Dolly the sheep, and a variety of other genetic 'breakthroughs' more rewarding in terms of instant publicity! Several international symposia on behavioural genetics have meanwhile made it clear that a combination of careful phenotype studies, animal models, gene mapping techniques and mathematical analysis should indeed make it possible to identify important genetic loci involved in human behavioural characteristics.

I should like to believe that this controversial issue has gone away, and that scientists in behavioural and cognitive genetics will remain free to pursue their research without medical pressures, distortion of facts or potential vilification of the workers involved. Alas, I fear this will not be the case, and that before long we shall see 'the IQ gene', however tenuous or minimal its role scientifically, taking its place on centre stage, with calls from politicians to use tests based on it in determining children's education (or to ban their use), from parents wanting to ensure that their own present and future children have such tests, and from a concerned public wondering how this research could have been going on without their knowing about it in advance. If and when that point is reached, there will not be much that scientists or clinicians can do to stop the consequent damage, and human genetics research in general may well be caught-up in the reaction. Our only safeguard is for this work, and other comparably controversial research areas, to be fully in the public domain from the outset and for us to do our best to ensure that people are as fully informed as possible about the potential dangers as well as the benefits of human genetics research, especially in such a sensitive area as behavioural genetics.

REFERENCES

1. Plomin R, McClearn GE, Smith DL *et al.* (1994) DNA markers associated with high versus low IQ: the IQ Quantitative Trait Loci (QTL) Project. *Behav. Genet.* **24**: 107–118.
2. Dooley T (1997) The testing questions. *MRC News*, **73**: 8–9.
3. Owen M (1997) Genetics: a new twist on psychiatry. *MRC News*, **73**: 10–15.
4. Medical Research Council (1997) *Genes and the Mind.* Medical Research Council, London.

Chapter

15

LIMITS TO GENETIC RESEARCH?

Human diversity, intelligence and race

A.J. Clarke

INTRODUCTION

The scope and pace of research in human genetics over the past 20 years has been astounding. At the time of my undergraduate degree in genetics – when Ed Southern published his famous paper on 'Southern' blotting in the *Journal of Molecular Biology* – mammalian genetics was regarded very much as the poor relation of the more rigorous and interesting genetics that was being practised more properly on the fruit fly and on fungi. I was not one of those with the foresight to predict the impending changes, but have observed developments since then with great interest.

The genetic basis of many Mendelian (single-gene) disorders in humans has now been elucidated, and even the common, but complex, multifactorial diseases are beginning to yield to genetic research, as already discussed in Chapter 7. The methods that have been developed to investigate diabetes and schizophrenia, however, may have a much broader application. Through identifying those portions of the genome that are shared by relatives who also share the complex trait being examined, it is possible to identify those regions of genetic material that influence the trait. This is not just true for diseases but also for physical features and even intelligence, personality traits and patterns of behaviour.

I do not doubt that a genetic basis will be found for some substantial part of the variation between individuals in respect of such non-disease characteristics; exactly how much of the variation is accounted for by genetic factors (i.e. how great is the heritability) will vary between populations and in relation to environmental factors. But should we be investigating these topics at all?

We would like to thank the Editor of the *British Medical Journal* for permission to base part of this chapter upon A. Clarke (1995) *British Medical Journal*, Vol. 311, pp. 35–38.

The technology developed to detect differences in DNA sequence between individuals can be applied to other contexts as well as to the genetic dissection of disease-associated and non-disease complex traits. Thus, differences between population groups can be identified and measured. In principle, this would allow a 'family tree' for mankind to be reconstructed, indicating the historical relationships between different populations. While this would intellectually be of great interest, it would inevitably lead towards the genetic study of *race*, and so the question arises once again: should we be using the tools of modern science to investigate this area?

What is in common between these two areas – the study of complex traits and of human genetic diversity – is that the tools of molecular genetics are being applied to examine normal human variation. Variation exists between individuals and between groups, and this research will measure this variation and draw inferences from it. The issues that are raised by such research overlap because both types of research touch on the subject of race.

NON-DISEASE TRAITS: INTELLIGENCE (IQ)

It is not possible to know in advance whether DNA technology will identify genetic variation associated with variation in intelligence quotient (IQ) and behavioural traits, but research projects along these lines are already in progress [1]. If a sufficient number of polymorphisms is examined in enough populations, then some such association may well be found; but what would this mean? It may be difficult and contentious to assign biological significance to such findings. While 'pure', 'disinterested', scientific curiosity may motivate this research, there are strong grounds for concern about the pursuit of such research and for doubting the practical wisdom of embarking upon it.

The short population history of mankind (the recent origins of modern man) make it likely that any genetic variation associated with variation in IQ will be non-randomly associated with other genetic polymorphisms in the same region of the genome, and it may be impossible to decide which of these polymorphisms is causally related to the variation in IQ. Indeed, the association may be misleading, having no causal relationship and either being the result of the pattern of sampling used in the study or having a complex relationship with social factors relating to the biological history of the population studied.

An almost inevitable consequence of finding an association between variation in IQ and genetic variation at one or more polymorphic sites will be the further finding that the allele frequencies at such sites (the frequencies of the different variants) will vary between different ethnic groups. Such variation between ethnic groups is found at virtually all polymorphic sites in the genome as a result of our species' population history. It is most unlikely that any such variation identified would be associated in an important way – causally – with differences between populations in IQ or other personality or behavioural traits, but it is highly likely that such differences would be widely publicized and misinterpreted. The allele associated with a higher IQ in one population will be found at a different frequency in any other population group studied. The biological significance of this is likely to be minimal, but the potential for misinterpretation or wilful abuse of the information is clear even if the original researchers pay no attention to ethnicity; other researchers are most likely to seek to replicate their findings in another population, or they may simply determine the allele frequencies in another population without even attempting to correlate these frequencies with IQ. When the 'favourable' set of markers is shown to be less frequent in one population than another, then this information will be wilfully misinterpreted to bolster the racist claim of the inferiority of that population.

Once genetic factors associated with variation in IQ have been identified, another set of likely problems will arise when institutions decide to test individuals to identify those with the highest IQ 'potential' in specific contexts, for example those with learning problems or those applying for employment or for higher education. Such applications of the results may be invalid and spurious, but could easily be introduced into practice nonetheless. The fact that a result obtained by DNA analysis may be of much less practical use than an examination result or even a simple IQ test would not hinder such DNA testing, because of the sense of legitimacy lent to the DNA testing by its origins in the precision world of genetics. The scope for destructive, self-fulfilling prophecies would be immense; those given a 'good' result would receive a superior education in the context of positive expectations, and those given a 'poor' result would be neglected.

Intelligence is such a highly valued human attribute that is so intimately linked with what makes us human that its genetic dissection will arouse very strong emotions in many people, including anger and fear. In addition to the potential harmful effects of promoting racism or the institutional misapplication of such research, therefore, the genetic dissection of intelligence implicitly but powerfully conveys two contentious messages:

(i) that people's genetic constitution *determines* their abilities and personal characteristics;
(ii) that knowledge of a person's IQ tells us something of central importance to that person's *worth* or value.

Researchers in this field will rightly repudiate such crude notions, but "actions speak louder than words"; whatever the personal views of those conducting the research, the research will of itself convey these messages.

The unwitting promotion of these views may have harmful effects on those with mental handicap and their families, quite apart from any strengthening of racism. By promoting the view that a person's worth is determined by his/her IQ, and that this is determined by the person's genetic constitution, the discrimination experienced by children and adults with mental handicap may be exacerbated, and the limited willingness of society to support such individuals and their families may be compromised. "What is the point of putting scarce public resources into the support or education of those who are biologically destined to contribute so little to society?", the argument might run.

Such applications of genetic pseudo-science to social policy are completely illegitimate, but are all too common; they have already been used earlier this century, not only in Germany but also in several other European countries and in the USA, and there are pressures now to repeat these errors. The extent to which genetic variation accounts for the overall level of variation in that trait (its heritability) tells us nothing about the likely response to educational or other social interventions. Indeed, the heritability of a trait is not a fixed, constant value but varies between populations and varies over time within a population: any estimate of heritability applies to a specific population at a specific time and in a particular environment. The misuse of heritability estimates to influence social policy inappropriately has been discussed fully by Rose, Lewontin, Alper, Beckwith, Billings and others [2–5].

Another potential problem arising from the genetic dissection of IQ relates to notions of 'reproductive responsibility'. Parents who have children with learning difficulties may be regarded as irresponsible, especially if children likely to have a low IQ can in the future be identified prenatally and hence potentially be aborted. But when is 'low IQ' enough of a problem to justify termination of an otherwise wanted pregnancy? One possible answer is, "when society makes it so", through intolerance, discrimination and a lack of solidarity with fellow citizens and fellow humans. I would argue that society already overvalues 'intelligence' as against other characteristics and qualities of persons, and that this research is likely to accentuate this still further.

Will couples seek prenatal testing for IQ with the intention of terminating pregnancies in which the fetus may not live up to their high expectations? Although this may not at first be a frequent request, I know that I do not want to be faced with it at all. A society that moves in the direction of terminating pregnancies because the fetus may become a child whose school performance would be merely mediocre is not one in which I could feel at home.

GENETICS, BEHAVIOUR AND SOCIETY

Crude genetic explanations of behaviour and personal differences strengthen the case of conservative thinkers (including sociobiologists, economists and political scientists) who argue that human society is inevitably hierarchical, competitive and male-dominated, and that 'free market' capitalism is the inevitable, best and final type of social organization. Francis Fukuyama, for example, argues that such a society represents the climax of evolution, after which there need be no further history. The rest of the world will come to resemble North America more and more closely and everything else will be mere detail [6].

While this ideology is clearly absurd when presented so crudely, it is also immensely powerful. It sustains the ambitions of world leaders, enabling men with power (and the occasional woman) to sleep peacefully in bed at night. The responsibility for any unfortunate results of their policies is removed from their shoulders, because their decisions can be seen as the inevitable consequences of the structural adjustments required by global evolution towards a single world market. They are therefore able to detach themselves from the practical results of their decisions, such as the devastating consequences in Africa, Asia or Latin America of high bank interest rates.

The appeal to genetics as a justification of social structures that exploit so many of our fellow humans amounts to a contemporary equivalent of the earlier and cruder social applications of evolutionary theory which allowed Victorian factory owners to treat their workers with such cruel indifference, and of the doctrines of race hygiene and racial superiority in Nazi ideology which allowed otherwise sensitive and intelligent professionals to consign their patients to death camps. Genetic research into IQ and personality may unintentionally bolster an understanding of genetics that is one element in the edifice of Western ideology, which allows politicians and others to justify their complicity in global economic exploitation.

Muller-Hill, the German geneticist who has worked so hard to reveal the close involvement of many German scientists and physicians in the Nazi regime, is similarly critical of the possible misinterpretation of such research. Despite this concern, however, he wishes to see it pursued in the hope that attitudes to genetic disadvantage will change and that society will eventually accept the truth wisely and compassionately, even if such knowledge is abused in the interim before wisdom, justice and compassion prevail [7]. While that view is understandable, my judgement is that the potential for abuse of such research should lead us to concentrate our research efforts elsewhere. There are many areas of medical and biological research that are more worthy of the time, energy and material resources than this.

Research into the genetic basis of non-disease traits, such as intelligence, should not be pursued until society has had a chance to discuss the ethical and social issues raised by such research, and to arrive at a consensus as to what information would be sufficiently important to risk the potentially harmful social consequences of such research.

THE HUMAN GENOME DIVERSITY PROJECT (HGDP)

The "Call for a worldwide survey of human genetic diversity: a vanishing opportunity for the human genome project" from Cavalli-Sforza and others [8,9] was inspired by the desire to sample and record the full range of modern human genetic diversity before the opportunity is lost through the processes of social integration and assimilation. The "Call" attempted to show sensitivity to the medical and social needs of threatened indigenous peoples, suggesting that research samples could be collected from such groups alongside efforts to meet these needs, but this suggestion did little to deflect the criticism that the "Call" encountered. Indeed, no attempt had been made to examine the type or validity of consent that would be sought in this undertaking, and the suggestion that some basic medical care could be offered by those collecting samples from indigenous peoples with no scientific understanding of the purposes of the project could have been understood as an attempt to seduce them into being co-operative. The "Call" made no mention of any long-term commitment to the well-being of the groups it would be studying, and its emphasis upon the 'vanishing opportunity' for the study even suggested that the authors had 'written off' the demise of some groups as inevitable. Given the contemporary extent of genocide, both as government policy in some countries and as the result of the inadequate control of land use in the face of demographic and commercial pressures in other countries, it is not surprising that such groups found the proposal to be offensive.

I had the opportunity to discuss these issues at a meeting organized by the European Science Foundation in 1995, and made extensive use of material provided by the Rural Advancement Foundation International [10] and the UNESCO International Bioethics Committee [11]. The paper by Margaret Lock was also most helpful, and I have assimilated a number of her ideas [12]. I am also particularly grateful to Darryl Macer for guiding me into the debate about the HGDP.

The vehemence of the objections made to the HGDP proposals will obviously have been exacerbated by the feeling of some indigenous peoples that decisions were being taken about them without consultation, especially when a target list of threatened groups was drawn up so that the collection of samples from them could be given priority. Being regarded simultaneously as a laboratory animal and a threatened species will inevitably lead to anger and resentment. In a wider context, the HGDP is also that part of the larger Human Genome Project which impacts most heavily on non-Western populations and on their systems of belief, which it may threaten. There are also Christian groups in Europe who have expressed concerns about science (i.e. genetic technology) invading the domain of the sacred, and who judge that our relationship to the earth should be one of stewardship rather than ownership, domination or control. They view the whole Human Genome Project as suspect, and the HGDP as an especially contentious element within that overall frame.

The rejection of the HGDP by some groups can also be viewed from another perspective, as a welcome development indicating a stage in the maturation of the self-awareness of indigenous communities who are learning to confront our contemporary global society. This must be respected, and once such groups have more self-confidence and less reason to fear their vulnerability to destruction by outside influences then they may be more willing to participate in this type of endeavour.

THE SCIENTIFIC VALIDITY OF 'RACE'

One set of challenges to the HGDP centres on the validity of concepts of 'race' or 'ethnicity' [13]. Whereas plotting the conjectured ancestral links between different species has some validity, it is much more questionable when similar methods are applied to trace the historical links between different human populations. The assumption that there have ever been discrete, isolated communities evolving independently is highly contentious, and one that social scientists and historians will not readily accept. Even now, population groups are defined in an arbitrary manner in relation to geographical or national boundaries or to patterns of language use, and do not represent coherent biological entities [14]. Population groups are cultural constructs that may have some use in everyday conversation but which represent such crude simplifications of reality that they become meaningless when asked to serve a more precisely defined conceptual role. There is the trend for descriptions of statistical variation between population groups to become descriptions of idealized racial types, and this ignores the processes of history and the mingling between groups that has happened worldwide and not just in the histories of Western peoples. 'Primitive', 'isolated' groups tend to be regarded as 'genetically pure'; the snapshot we have of how such groups now live is regarded as an adequate description of how they have always lived. This ignores these groups' own particular histories and is therefore bad science as well as unacceptably ethnocentric. It is too easy to regard history as linear rather than like a mesh, especially when the methods of cladistics (the study of relationships between species, where interbreeding is not possible) are borrowed for studies within a species.

FEAR OF PROMOTING RACISM AND GENETIC DISCRIMINATION

To rebut the criticism that the HGDP will promote racism, it is argued that genetic differences within populations are generally greater than those between populations, and that there is no evidence that any one ethnic group is more gifted than another. But this very argument assumes that races do exist, and that it would be intelligible to search for such differences – that they could be found. Given the search for IQ genes, aggression genes, sexuality genes, etc., it is only a matter of time before IQ-associated polymorphisms (or similar) are studied in different populations, leading to entirely predictable misunderstandings that will receive much more public attention than the cautious interpretation of these findings made by the scientists involved will ever do.

This argument, that facts can counter racism, presupposes that racism is a rational position adopted in error and amenable to reasoned argument; in fact, it is a prejudice – a deep-seated attitude – which will twist facts to its own ends. A decision for or against prejudice is fundamentally a value judgement – a moral choice, an existential decision – and the only effective arguments against such prejudices are moral arguments relating to values.

Genetic discrimination and prejudice is not only based on notions of race, however, but also on an exaggerated sense of the importance of intelligence and certain aspects of physical and mental health, on an exaggerated sense of the contribution of inheritance to health and well-being, and on an exaggerated sense of the importance of group welfare as opposed to the individual. Racist attitudes are thus a component of a more general eugenicist mindset. Scientific knowledge will not undermine racism, or other discriminatory, eugenic practices, and may be perverted to support them. There are only too many examples of scientists and medical practitioners who have carried out unethical research or developed theories that have promoted prejudice and unjust discrimination.

INFORMED CONSENT FOR HGDP RESEARCH?

Even in Western Europe and North America, difficult issues arise in the context of the informed consent required for medical research, including genetic research. These issues will inevitably be much more complex when researchers attempt to obtain biological research samples from the members of indigenous groups essentially naïve to the nature and purposes of Western science. How valid is the consent of Western-educated representatives of these peoples? Can third-party mediation be used between the community and the researchers? Does a kinship or community have a right to genetic privacy? Before consent to participate in such research could be regarded as valid, detailed discussions would be required concerning the timescale for information-giving and discussion, the documentation of consent to participate and of decisions not to participate, and for the resolution of differences that may arise within a community in response to the research and which could be socially damaging even if the invitation to participate in the research is made in a very sensitive manner.

Can consent to participate in HGDP research be regarded as valid if the research teams offer medical or other tangible benefits to individuals or communities that do participate, or if they are seen to work closely with those providing such services? Would such consent not be subject to coercion? One model for how genetics inquiry can be located within a broader framework of interest in the indigenous peoples has been provided by the 'Expedicion Humana' of Javeriana University in Bogota in relation to many indigenous communities in Colombia.

What should happen to those samples already collected without such planning and without full attention to consent procedures? If they are used by the HGDP, then does that not make a mockery of the project's respect for consent?

EXPERIENCES OF EXPLOITATION

Another obstacle to the HGDP is the long list of previous ways in which indigenous peoples have suffered from outside interference. The HGDP can be seen as the latest in a long series of episodes of exploitation, and it is not surprising that the responses of some indigenous peoples have been suspicious or even hostile. These previous experiences have included conquest, militarization, human rights abuses and the exploitation of natural resources and of human labour. Whether perpetrated by colonial powers, a remote national government or commercial interests, these three forms of interference from without can be hard to distinguish, and they have generally led to benefits for outsiders and difficulties for the indigenous people.

The search for genetic riches in the South (the Third and Fourth Worlds) has previously concentrated on plants, and has led to the commercial exploitation of countries that donated the source genetic material. This has been a 20th century sequel to the earlier Western exploitation of tobacco, the tomato, maize, the potato and rubber. How rich the indigenous peoples would now be if they had patented these crops! Now, it may be thought, it is the turn of the vanishing human resources of these countries to have their genetic potential exploited.

Participants in the HGDP are likely to gain little from the project, and will suffer intrusion into their lives and their ways of life. It may be acceptable to invite their participation, however, because of the intellectual interest and the potential practical benefits of the HGDP as long as certain conditions are met. The foremost of these conditions is to give credible reassurances that participants will not look back in 50 years and realize that they had been exploited once again.

FEARS OF COMMERCIAL EXPLOITATION

Just as the academic discipline of social anthropology served the British Empire well in the past, linking with the colonial authorities to their mutual benefit, so the respectable, academic enterprise of genetics research forges links with modern biotechnology companies. In both cases, the objects of academic study are open to exploitation.

With commercial interests controlling the direction of so much basic research in human genetics, and especially the application of human genetics to medicine, there is a legitimate fear that royalties on genetic discoveries enabled or facilitated by the HGDP would become a tax on health care – amounting to a fresh exploitation of the already disadvantaged South of the world, another means of transferring resources from the impoverished South to the industrialized North.

It is entirely understandable that indigenous peoples and their representatives might expect corporations and governments to gain financially from the knowledge that emerges from the study of peoples whose very existence is threatened by those same corporations and governments, and who will gain nothing from co-operating with the HGDP. The denunciation of intellectual property rights, patent law and the apparatus of informed consent as tools of legalized Western deception, exploitation and blackmail is similarly understandable. Even the HGDP's proposed solution – that the HGDP should set up local research laboratories in each region of the globe to promote a global sense of ownership of the project – could generate problems by diverting foreign aid budgets away from providing concrete help to needy people.

A possible resolution of the problems in international development may come from the global North stepping back from such close involvement in the South; it would help if the North stopped being a problem for the South rather than compounding their difficulties by interfering more and more. In the case of social anthropology, anxious to distance itself from its historic involvement with colonialism, there has been a move towards studying advanced, industrialized societies. Perhaps the HGDP could learn from this by targeting its efforts predominantly on the industrialized societies of Europe, North America and East Asia. In these countries, there would be fewer issues of research ethics because the level of informed consent that could be achieved would be much higher than with many of the indigenous peoples. If these studies proceeded smoothly and were intellectually productive, then it would become easier to seek the participation of groups from other parts of the world who live in different circumstances.

CONCLUSIONS

To avoid future charges of exploitation, the HGDP should acknowledge the reasonableness of the fears expressed in relation to the commercialization of genetic research. While fears of genetic data being used in biological warfare against indigenous peoples may not be scientifically credible – the genetic differences between neighbouring, warring peoples are likely to be far too small to permit any such use – the reasons why these fears have arisen must be respected.

The focus of HGDP on population groups that may vanish, by assimilation or by extermination, should be replaced by less threatening studies carried out in regions where the populations have few fears about the project, and where the validity of the participants' informed consent is less in doubt. This would involve fewer studies of the grand, global evolution of mankind and more micro-evolutionary studies of genetic variation within populations and on the processes of population change and mixing through recent history rather than the millennia of prehistory.

The subjects of HGDP research should become participants in research, actively involved in a dialogue with HGDP rather than mere objects of study.

If some studies do not happen because of this, then that is the (rather small) price that must be paid for treating fellow humans with respect.

The HGDP must not seduce participants with promises of benefits and financial gain to individuals or to communities. The dialogue with potential research participants among the indigenous peoples should include a detailed discussion of the most appropriate ways of gaining valid, informed consent for participation both from the individuals involved and their communities.

The HGDP should divorce itself from commercial influences and should firmly reject the patenting of discoveries throughout the Human Genome Project as a whole, arguing that specific gene sequences identified through studies of particular population groups should not be patented. Individuals suffering from a wide range of genetic diseases have co-operated with researchers in human genetics over many years, in the hope that their contributions to the research would help to understand and then to combat the diseases found in their families. They and other members of their families have been betrayed by the genetics research community now that the findings of the research that they made possible have been patented and the applications of the new knowledge have been commercialized. What has happened with the sufferers of genetic diseases should not also happen with those indigenous peoples who decide to co-operate with the HGDP.

REFERENCES

1. Plomin R, McClearn GE, Smith DL *et al.* (1994) DNA markers associated with high versus low IQ: the IQ quantitative trait loci (QTL) project. *Behav. Genet.* **24**: 107–118.
2. Rose S, Kamin LJ, Lewontin RC (1984) *Not in Our Genes: Biology, Ideology and Human Nature*. Penguin Books, Harmondsworth.
3. Billings PR, Beckwith J, Alper JS (1992) The genetic analysis of human behaviour: a new era? *Soc. Sci. Med.* **35**: 227–238.
4. Lewontin RC (1993) *The Doctrine of DNA: Biology as Ideology*. Penguin Books, Harmondsworth.
5. Alper JS, Beckwith J (1994) Genetic fatalism and social policy: the implications of behaviour genetics research. *Yale J. Biol. Med.* **66**: 511–524.
6. Fukuyama F (1992) *The End of History and The Last Man*. Penguin Books, Harmondsworth.
7. Muller-Hill B (1993) The shadow of genetic injustice. *Nature*, **362**: 491–492.
8. Cavalli-Sforza LL, Wilson AC, Cantor CR, Cook-Deegan RM, King M-C (1991) Call for a worldwide survey of human genetic diversity: a vanishing opportunity for the Human Genome Project. *Genomics*, **11**: 490–491.
9. Bowcock A, Cavalli-Sforza LL (1991) The study of variation in the human genome. *Genomics*, **11**: 491–498.
10. RAFI (1993). *Patents, Indigenous Peoples, and Human Genetic Diversity*. RAFI Communique, May 1993. Ottawa, Canada.
11. UNESCO International Bioethics Committee (1995) *Report of Population Genetics Subcommittee*. UNESCO.

12. Lock M (1994) Interrogating the human diversity genome project. *Soc. Sci. Med.* **39**: 603–606.
13. Macbeth H (1997) What is an ethnic group? – a biological perspective. In: *Culture, Kinship and Genes* (eds A. Clarke, E.P. Parsons). Macmillan, London.
14. Marks J (1995) *Human Biodiversity: Genes, Race and History*. Aldine de Gruyter, New York.

Section IV

THE ABUSE OF GENETICS, PAST, PRESENT AND FUTURE

Workers in all fields of science and medicine like to feel that they and their colleagues have contributed in some degree to the benefit of mankind, whether by some practical advance involving human health, or by increasing the pool of fundamental knowledge on which all advances are ultimately based. We are aware that this knowledge can sometimes be used by others for harmful purposes, but this is rarely something that can be directly controlled or avoided by our own actions. Some scientific fields have traditionally been uncomfortably close to these harmful applications, such as physics and microbiology in relation to atomic and microbial warfare. Genetics is another such area, linked with controversy and potential misuse since its earliest years.

The three chapters in this section cover widely different topics, but all can fairly be grouped under the broad category 'abuse of genetics', either actual or potential. The first two have a common theme in Nazi Germany, illustrating that scientific and medical abuse are much more likely to occur when a particular social system encourages and institutionalizes the practices. These historical lessons from the past are highly relevant to the present and future, and concern is rightly heightened by the hugely increased power of modern genetics since the time of the Third Reich. Huntington's disease, the focus of the second chapter of the group, is a notable example of how molecular analysis could be misused if the social conditions of Nazi Germany were to occur again today.

The final chapter, dealing with the return of 'eugenics' in the form of genetic legislation in China, again raises serious questions for the applications of genetics to medicine in the present and future, not just in China, but worldwide. We do not yet know to what extent the law will be applied, nor whether it will lead to actual abuse; the topic remains one of controversy both in China and outside. Perhaps it is appropriate that this book should end on a note that is inconclusive, and that reminds us that the future of genetics and the way in which it is used or misused, is to a large extent in our own hands. Whether we are clinicians, scientists, or members of the concerned public, we need to work actively to ensure that our views are heard and that genetics is used for human good – not to cause harm.

Chapter
16

THE NAMING OF GENETIC SYNDROMES AND UNETHICAL ACTIVITIES

The case of Hallervorden and Spatz

P.S. Harper

Many rare disorders, particularly those of a genetic nature, receive eponymous titles related to the original describers of the condition. While such titles have their disadvantages, it is likely that many will persist as a result of long tradition and lack of a generally agreed alternative; medical students and others frequently remember the name even when unable to recall anything else about the disorder.

Few people using an eponymous title for a syndrome will know, or wish to know, details of the individuals whose name the disorder bears, though for the more famous (e.g. Duchenne, Huntington), considerable biographical information may be available. Several works [1,2] have attempted to provide this information in a more general way on grounds of historical interest.

At first sight, it does not seem wise to question whether ethical considerations, rather than priority, should determine whether a person's name is attached to a disorder that they have described. However, there may be instances where the ethical issues are so overwhelming and so directly related to the work concerned, that it would be wrong to use the name of such investigators. I consider the case of Hallervorden and Spatz to be such an example.

The book *Murderous Science* by Professor Benno Muller-Hill [3], documents the involvement of numerous clinicians and scientists in the mass killings of handicapped or mentally ill patients in Nazi Germany. The English translation, published in 1988, had a profound impact on my views of this period. Before reading it, I had, like many others, considered that scientists and medical staff were largely passive agents in the process, whereas Muller-Hill's

Part of this chapter is reproduced from the *Lancet* (1996) by kind permission.

book indicates that many used the killings to further actively their own research and careers.

Prominent among such men were Professors Hallervorden and Spatz, head of Neuropathology and Director respectively of the prestigious Kaiser Wilhelm Institute of Brain Research. In 1922, Hallervorden and Spatz had described a new form of familial brain degeneration [4], occurring in five of a sibship of 12, and characterized by unusual histological features, notably iron deposition. The name 'Hallervorden–Spatz disease' or syndrome has been generally used in the subsequent literature. Although this description dates from before the Nazi period and there is no suggestion that the brains studied were obtained in an improper way, the subsequent work of these investigators was very different.

Both Hallervorden and Spatz were closely associated with the Nazi extermination policies and the nearby mental hospital functioned as an extermination centre. Hallervorden is stated to have visited the centre, provided it with a technician, and on one occasion to have personally removed the victim's brain. Muller-Hill quotes his statement to the American interrogating officers at the end of the war:

> I heard that they were going to do that and so I went up to them: "Look here now, boys, if you are going to kill all these people at least take the brains out, so that the material could be utilized." They asked me: "How many can you examine?" And so I told them an unlimited number – "the more the better".' I gave them fixatives, jars and boxes, and instructions for removing and fixing the brains and they came bringing them like the delivery van from the furniture company. There was wonderful material among those brains, beautiful mental defectives, malformations and early infantile diseases.

There is no doubt that this opportunity benefited both Hallervorden's research and the reputation of the Institute as a whole.

> This material is constantly being added to by the post-mortem department of the mental hospital in Gorden, which is directed by Dr Eicke, who is also an assistant at this institute. All the cases examined there are investigated further by me and a written report is deposited. In addition, during the course of this summer, I have been able to dissect 500 brains from feeble-minded individuals and to prepare them for examination.

The direct role of Spatz is less clear; perhaps as Director of the Institute he did not choose to enquire closely into the details of a specific department's work. Be this as it may, both he and Hallervorden were reinstated in their positions as Director and department head in 1948, continuing until their retirement and dying in 1965 and 1969 respectively.

It might be thought that in the light of this and other information, the names of Hallervorden and Spatz would have been removed from the disease that they described. Yet there is no mention of these abuses in subsequent papers and reviews of the disorder, whether of the genetics [5] or the neuropathology [6]. The standard German textbook on clinical syndromes [7], published in 1981, notes as biographical information simply: 'Hallervorden, Julius. Deutsche Neurologe, Giessen, 1882–1965. Spatz, Hugo, Zeitgenoss deutscher Neuropathologe, Frankfurt a. Main.' A more recent work on medical eponyms [8] notes that Hallervorden 'loved symphonic work' while Spatz is cited as 'a very humane and thoughtful person'. There is no mention of the relationship of their work to the killing of patients in these accounts, or in obituaries of Hallervorden [9–10]. Hallervorden's activities are mentioned, however, in a memorial ceremony held for victims in 1990, the transcript of which was published in a neuropathology journal [11].

In 1988, at a genetics congress where a lecture was given on eponymous syndromes, I raised my concerns about Hallervorden and Spatz, and was criticized for introducing political issues into science. I had no cause to raise the topic again until 1995, when the proband of a family with an undiagnosed neurodegeneration whom I had seen repeatedly for genetic counselling over many years, was found on post-mortem to have 'Hallervorden–Spatz disease', awakening and focusing my previous concerns.

Most families find it extremely helpful to be able to give a specific name to a disorder, but in this instance, I felt unable to give them the name, let alone the information described above. In my mind was the distress they would feel if the memory of their dead child was to be connected with such criminal acts. I have not yet found a satisfactory alternative name, except for the more generic group of 'neuroaxonal dystrophy'. In writing this, I realize that I may cause distress to others who have already been accustomed to the name of 'Hallervorden–Spatz disease' in their family. It is even possible that the name might be used for a lay support society, as is commonly formed by families with rare disorders.

It could be argued that once ethical considerations are introduced into medical nomenclature, the process could be extended without limit; for example many pathologists have based descriptions on aborted fetal material, which some might consider unethical. My personal view is that the crimes of the Nazi era, and the involvement of clinicians and scientists in the deaths of the sick and handicapped and of racial minorities, are of such enormity as to make it indefensible to use the names of those involved for disorders, even though their original descriptions may have involved no wrongdoing. When a specific molecular defect is found, a logical scientific name should be possible to agree on; meanwhile we should be content to use the term 'neuroaxonal dystrophy', to abandon the term 'Hallervorden–Spatz disease', and to consign it to an episode of history that is behind us but hopefully, never to be forgotten.

POSTSCRIPT

Before the above paper was published in the *Lancet*, I met with the parents of the affected children in the family with whom I had been personally involved, gave them a copy, and told them frankly of the issue. As I expected, they were deeply shocked and totally surprised. As the father said: "we expect these to have been wise and humane people".

The paper itself produced interesting and moving correspondence, some published in the *Lancet* [12,13], illustrating the profound effects the Holocaust continues to have on survivors and others. With a single exception, the response was strongly favourable, both of the paper having been published and of the proposal to use a different name for the condition. Several authors of textbooks wrote to say that they planned to make this change, while my attention was drawn to an article that I had overlooked [14], with associated correspondence also supporting a change of name. An interesting account has also been given [15] of the American Psychiatrist Leo Alexander, seconded from the US military to undertake the interviews from which quotations were given in my article. This report was largely suppressed – or at least not made public – until recently, illustrating how all parties involved had a vested interest in these uncomfortable facts remaining hidden. Only now are the full details on this whole era beginning to emerge, with implications for the present and the future that we must not ignore.

REFERENCES

1. Beighton P, Beighton G (1991) *The Man Behind the Syndrome*. Springer, Berlin.
2. Jablonski S (1991) *Jablonski's Dictionary of Syndrome and Eponymic Diseases*. Krieger, Malabar, FL.
3. Muller-Hill B (1988) *Murderous Science*. Oxford University Press, Oxford.
4. Hallervorden J, Spatz H (1922) Eigenartige Erkrankung im extrapyramidalen system mit besonderer Beteiligung des Globus pallidus und der Substantia nigra. Ein Beitrag zu den Beziehungen zwischen diesen beiden Zentren. *A. Ges. Neurol. Psychiat.* **79**: 254–302.
5. Elejalde BR, Meredes de Elejalde M, Lopez F (1979) Hallervorden–Spatz disease. *Clin. Genet.* **16**: 1–18.
6. Adams JH, Duchen LW (eds) (1992) *Greenfield's Neuropathology*, pp. 857–859. Edward Arnold, London.
7. Leiber B, Olbrich G (1981) *Die Klinische Syndrome*. Urban and Schwarzenberg, Munich.
8. Firkin BG, Whitworth JA (1987) Hallervorden–Spatz disease or syndrome. In: *Dictionary of Medical Eponyms*, pp. 214–215. Glaxo, London.
9. Spatz H (1966) Erinnerungen an Julius Hallervorden (1882–1965). *Nervenartst*, **37**: 477–482.
10. Van Bogaert L (1967) Obituary. Julius Hallervorden (1882–1965). *J. Neurol. Sci.* **5**: 190–191.

11. Peiffer J (1991) Neuropathology in the Third Reich. Memorial to those victims of National-Socialist atrocities in Germany who were used by medical science. *Brain Pathol.* **1**: 125–131.
12. Shevell M (1996) Naming of syndromes. *Lancet*, **348**: 1662.
13. Jacoby R, Oppenheimer C (1996) Naming of syndromes. *Lancet*, **348**: 1662.
14. Shevell M (1992) Racial hygiene, active euthanasia and Julius Hallervorden. *Neurology*, **42**: 2214–2219.
15. Shevell M (1996) Neurology's witness to history: the combined intelligence operative sub-committee reports of Leo Alexander. *Neurology*, **47**: 1096–1103.

Chapter 17

HUNTINGTON'S DISEASE AND THE ABUSE OF GENETICS

P.S. Harper

The paper forming the core of this chapter[a], written 2 years before the isolation of the Huntington's disease (HD) gene, was a particularly uncomfortable one to write, involving a disease that had been my own particular interest for many years. It also showed me how ignorant I had been for most of this time regarding the abuse of genetics in this condition. I had known of the work of Panse in Germany and Davenport in America, and had frequently cited their research, but had been completely unaware of their direct and specific involvement in the development of eugenic legislation and, in Nazi Germany, the compulsory sterilization and later the killing of HD patients. I might well have remained in this state of ignorance had not my editorial responsibility for *Journal of Medical Genetics* meant that a steady flow of general books on genetic topics came my way to arrange reviews, many of which I was able to read before they were sent on to their intended reviewer.

Among these books were several general accounts of the eugenics movement, covering both America and Europe, notably those of Kevles [1], Adams [2] and Mazumdar [3], while that of Weindling [4] covered the Nazi period in depth. Undoubtedly, though, it was Muller-Hill's book *Murderous Science*, published in English translation in 1988 [5], that had the greatest impact, even though its author would make no claim for it as a work of historical scholarship. This book should be compulsory reading, and deeply disturbing reading it is, for all clinicians and scientists working or training in human genetics. I remain surprised, given the controversy caused by the book in Germany, how few people, in other countries have read it or are aware of the full story of the involvement of their predecessors in this terrible and tragic period. At a time when neuropsychiatric and behavioural genetics are again entering controversial areas,

[a]Reproduced from the *American Journal of Human Genetics* by kind permission of The University of Chicago Press (© 1992 by the American Society of Human Genetics. All rights reserved).

everyone involved should be fully aware of the long shadow that is still cast by the abuse of genetics in these disorders. While I have written here only about HD, my own field of expertise, similar accounts of abuse could be given for schizophrenia or other mental disorders.

HD, with its combination of progressive neurological and mental deterioration in adult life and with its autosomal dominant transmission from generation to generation, provides particularly difficult problems in genetic counselling, as well as causing a severe burden for the families in which it occurs [6]. It has been the subject of genetic studies since the first recognition of its Mendelian nature [7,8], and in many ways it provides a model for other late-onset neurodegenerative disorders.

Since the discovery of a closely linked genetic marker for HD [9] and since the exclusion of multilocus heterogeneity [10], presymptomatic detection for HD has been feasible and, with the advent of even closer markers, has become widely used in a few major centres [11,12]. These applications, together with the very real practical and ethical problems encountered in presymptomatic detection of HD [13], have led to fears that the new techniques of molecular genetic prediction could be abused in testing of HD and for other, comparable disorders, where similar approaches will become possible. Most of this concern relates to the future; it is perhaps less well recognized that abuse of genetics has already occurred in the past in relation to HD. In drawing attention to this abuse, I hope that workers in medical genetics, clinicians involved with the management of HD families, and those engaged in work on other progressive genetic disorders will be sensitized to the issues and potential dangers involved.

The current review arose from the collection of historical and other material on HD, for the preparation of a book [6]. No systematic search was made for evidence of abuse in different countries, so it is quite possible that problems have not been confined to those countries (Germany and the USA) for which published material was readily available and which have been the principal focus for historical research regarding eugenics. I am aware that this approach has been a superficial one, and it is likely that unpublished data exist that might give a more complete picture of the subject.

HD AND THE EARLY EUGENICS MOVEMENT

The first extensive genetic study of HD was carried out in America by Charles Davenport, with the assistance of Elisabeth Muncey and was published in 1916 [14]. Davenport's recognition of the Mendelian dominant inheritance of the disorder is clearly laid out in his earlier monograph [15], in which his general views on eugenic measures are also stated in relation to numerous other medical and social diseases and characteristics, many of which are now recognized to have no clear Mendelian basis. Davenport carried out a major survey of the extent and origin of the HD families in New England and concluded that they were descended from a very small number of founding individuals who had emigrated from England. His views on this are clearly stated: "All these evils in our study trace back to some half-dozen individuals including three brothers, who migrated to this country during the 17th century. Had these half-dozen individuals been kept out of this country much of misery might have been saved" [14]. Davenport's conviction that HD in America might largely have been avoided by the careful screening of immigrants may well have been an influence in the development of subsequent restrictive immigration laws in the United States, in which his colleague Laughlin played a major role [1]. It is not clear, however, that a family history of HD was ever the ground for refusing an individual immigrant, nor is it likely that such a policy would have ever been effective in preventing HD. Subsequently, American and other population studies have clearly shown the multiple origins of HD and the variety of ethnic groups from which it has been derived [16].

The screening of immigrants was not the only eugenic policy advocated by Davenport in relation to HD. He clearly (and correctly) recognized that most patients did not represent new mutations but had received the gene from an affected parent, and he stated in characteristically blunt terms his views as to the remedy:

> It would be a work of far-seeing philanthropy to sterilize all those in which chronic chorea had already developed and to secure that such of their offspring as show prematurely its symptoms shall not reproduce. It is for the state to investigate every case of Huntington's chorea that appears and to concern itself with all the progeny of such. That is the least the state can do to fulfil its duty toward the as yet unborn. A state that knows who are its choreics and knows that of the children of every one of such will (on the average) become choreic and does not do the obvious thing to prevent the spread of this dire inheritable disease is impotent, stupid and blind and invites disaster. We think only of personal liberty and forget the rights and liberties of the unborn of whom that state is the sole protector. Unfortunate the nation when the state declines to fulfil this duty. (Ref. 14, p. 215.)

This passage epitomizes the inherent tension – and, in some situations, conflict – between the rights of the individual and those of the state; it highlights also the potential conflict between the public health aims relating to population prevention of a disorder such as HD and the individual genetic counselling of family members that has as its primary tenet a respect for the decisions and autonomy of those counselled (see Chapter 10). The role of the state in reproductive decisions is set out uncompromisingly and foreshadows later developments in Germany, which will be discussed below. In one respect Davenport misses the most important point on which any population prevention of HD must rest: most HD genes are transmitted not by affected patients but by healthy individuals at risk for the disorder; thus any effective eugenic policy of this nature would have had to legislate for the compulsory sterilization of those at risk.

Lest it be thought that these views were confined to America, it should be noted that an early British study, that of Spillane and Phillips [17] in south Wales, had a comparable view on the topic: "Perhaps with repeated advice and education, some would voluntarily abstain from marriage, but the majority would no doubt be prepared to accept the even chance that nature offers them. We are thus left with the conclusion that only legislative measures will eventually succeed in eradicating the disease" [17].

A more thoughtful and farsighted view was expressed around the same time, in the British study by Julia Bell:

> The almost continuous anxiety of unaffected members of these families over so long a period must be a great strain and handicap, even if they remain free from disquieting symptoms; it is thus of urgent importance that some means should be sought by which the immunity of an individual could be predicted early in life, both from the point of view of relief to those who carry no liability to the disease and as an indication to others that they should abstain from parenthood. No facts in the clinical histories of patients provide definite guidance in this matter prior to the onset of symptoms, but the development of the science of genetics may at some future date enable us to obtain information concerning the inherent characteristics in such cases. [18]

Compulsory sterilization laws were not introduced in Britain, but they were in the USA, though their use was principally applied to the mentally handicapped and to those with criminal tendencies. In the case of HD, the principal application of the eugenic views quoted above was to come in Nazi Germany.

The involvement of patients with genetic and psychiatric disorders in the race-hygiene policies of Nazi Germany is well-known, but perhaps less recognized are the extensive background of eugenic views existing well before the Nazi period in Germany and the international influences that helped these policies to develop. The books of Kevles [1] and Weindling [4] give a detailed picture of how these developments reach back even into the previous century, of how scientists and physicians such as Fischer, von Verschuer, and others, formulated the ideas underlying subsequent policies applied by the state authorities, and of how Davenport and other eugenicists were directly involved with and supportive of developments in Germany.

HD, with its combination of mental involvement and inherited nature, was an obvious candidate for inclusion in any restrictive legislation; it is specifically listed as one of the nine categories of disorder suitable for compulsory sterilization in the act of 14 July 1933 [5]. Only affected individuals were included; family members were to be closely supervised, especially those with 'anomalies of character'.

Some estimate of the total number of HD patients actually sterilized compulsorily can be gained by the indications given in a doctoral thesis which cites nine HD cases among 950 sterilizations in Leipzig [19]. On the basis of

350 000–400 000 compulsory sterilizations undertaken as a whole, B. Muller-Hill (personal communication) has suggested that there could have been 3000–3500 sterilizations undertaken in which HD was the indication.

Compulsory sterilization was the prelude to the killing of those with genetic disorders, mental handicap and psychiatric illness, along with racial minority groups. Meyer (1988) has reviewed the infamous 'T4' extermination policy, especially in relation to psychiatric disorders, and notes that HD was specifically included in the list of those conditions for which killing was indicated. No accurate data exist on how many HD patients were killed in this program, nor have I so far been able to obtain from living family members any firsthand information on instances that occurred in their families. It is possible that members of the lay HD societies in Germany (societies that were legalized in former East Germany only shortly before reunification) may possess such information. In an article in the UK Huntington's Disease association newsletter, Hjardeng [20] has interviewed a number of those involved and has suggested that over 100 HD patients may have been killed at one psychiatric unit in 1941 alone.

The way in which compulsory sterilization and other abuses were administered has been described by Weindling [4], who emphasizes the predominant role of professionals in the process: "A system of hereditary courts was established; each tribunal was composed of a lawyer, a medical officer and a doctor with specialist training in racial hygiene. The medical officer could initiate proceedings as well as adjudicate, and doctors were in the majority. The state established primacy over reproduction, but left the operating of the controls to the medical profession" [4]. In the case of HD this facade of legality caused the sterilization of some Leipzig patients to be rejected on grounds of uncertain diagnosis, while in Dusseldorf it was, in some cases, refused on grounds of advanced age or deteriorated general health [21].

The role of those scientists and clinicians specifically working on HD in Germany during this period is particularly important to document. Two such workers were Friedrich Panse (1899–1973), professor at the Psychiatric Neurological Research Institute in Bonn, and Kurt Pohlisch (1893–1955) psychiatrist and director of the same Institute. Panse was one of the most distinguished workers in the field of HD at that time; he undertook a detailed study of the disorder in the Rhineland, and his monograph, *Die Erbchorea* [22], published in 1942, contains a wealth of detailed and accurate genetic analysis which has been extensively quoted in the HD genetic literature. Less well known is Panse's involvement in the genetic abuse of this period; both he and Pohlisch acted as expert witnesses in relation to the Genetic Health Courts mentioned above, including the extermination of psychiatric patients [5]. Both were Nazi Party members and were closely involved in the drawing up of the race-hygiene laws [5]. Panse's role in reporting the HD cases and relatives who had formed part of his study is clearly stated in his book: "We proceeded in a manner that we reported all choreic cases, and moreover all suspicious cases and finally all not yet choreic sibs and offspring as being at risk to the health authorities . . . 79 cases located and diagnosed by us were reported to the health administration. They have been passed on to the Genetics Health procedure, if they were of an age to procreate" (ref. 22, pp. 232–233). There can be no doubt that Panse must have been fully aware of the likely fate in store for those individuals whom he reported to the Genetic Health courts; the conflict of interest between the perceived public health role of the geneticist and duty to the individual is clearly highlighted by this situation, as are the dangers of misuse of research information and information held on genetic registers.

Despite such involvement, after the war both Panse and Pohlisch were acquitted of criminal acts and were reappointed to senior academic posts. Both the widespread denial that abuses had occurred and the reintegration of such individuals into the medical scientific hierarchy are vividly illustrated in the book by Muller-Hill [5] and are likely to have been major factors in the current mistrust of new genetic developments that is widespread in Germany today. It is striking that subsequent work on HD in Germany makes almost no reference to the fate of HD families under the Third Reich; the 1972 monograph of Wendt and Drohm [23], bringing together work done by Wendt and colleagues over the previous two decades, notes only that, for the period 1933–1945, "it was looked on as a disgrace to have a genetic disease". Likewise, a recent article, 'Ethics and medical genetics in the Federal Republic of Germany' [24], raises concerns about molecular testing for HD but says little about past abuses.

The saga of the Panse HD families is not yet over, for there has been in Germany considerable debate as to whether they should be used for current research and for tracing families. Panse's full collection of data on HD families, along with data on individuals with other genetic disorders, was contained on file cards, essentially representing a genetic register. Towards the end of the war these cards, several million in number, were transported in three trainloads into what subsequently became East Germany. They remained there until they were recently located and brought back to form the foundations of a new register, a step that aroused further major controversy, as reported in the periodical Stern [25]. There is currently a legal moratorium over their use, until the ethical and practical issues are resolved.

ABUSE IN THE PRESENT AND FUTURE

While the kind of systematic and deliberate abuse that Nazi Germany practised on patients and families with HD (or other genetic disorders) is difficult to envisage occurring again at the present time, it would be most unwise to discount the possibility, particularly when one considers the power inherent in new developments in genetics. Two such developments: (i) the existence of molecular tests allowing prediction of the HD gene, and (ii) the feasibility of keeping computerized genetic registers, could both be the source of serious abuse and would undoubtedly have been used in this way had they existed at the time of the Third Reich.

Inappropriate use of molecular predictive tests has already been documented in several centres (see Chapter 3); examples in our own series [26] include requests for testing individuals in prison or compulsorily detained in psychiatric hospitals, as well as requests both for the results of tests in relation to health and life insurance and for the testing of young children and infants being placed for adoption [27]. While such inappropriate referrals have usually been well intentioned, the potential harm that could be caused is considerable. The simplification of testing resulting from identification of a specific mutation for HD will greatly increase the likelihood of such misuse unless testing can be closely regulated and monitored.

The development of genetic registers is probably an even greater cause for concern, since here the potential is for abuse of information on an entire population of HD patients and relatives. Inadvertent disclosure of such sensitive information could cause considerable harm to the employment and insurance prospects of individuals on the register. The power of current microcomputers means that an immense amount of information can be stored and easily accessed, necessitating the strictest security for genetic registers. Paradoxically, however, such security may be easier to achieve, since such a register can be kept confined within a single department, without the necessity of transferring and transporting sensitive identifying data in a way that could lead to either loss or abuse. Most countries, including Britain but not yet the USA, now have legislation that regulates the use of computerized databases of this type; in the British model this necessitates registration of the register itself and allows access by an individual to his or her recorded personal information.

HD is one of the most important genetic disorders of later life, but it can also serve as a valuable model for those numerous other neurological degenerations in which similar ethical issues and potential problems can arise in relation to predictive tests and genetic counselling. The issues relating to HIV status and testing are also closely related [28,29], raising comparable problems of confidentiality and need for privacy. The abuses of the past in relation to HD can provide important lessons for this whole group of disorders, as can our current experience of genetic prediction, and they are given added relevance by the power of the new molecular approaches and of computerized genetic registers.

Finally, it would be unwise to assume that the abuses described here will not occur again in our present social system. This review has shown that they have not been confined to totalitarian societies and that geneticists and clinicians directly involved with HD have been prominent among those responsible both for the abuses and for the policies underlying them. It is thus essential that we be prepared to recognize what has happened in the past, if we are to avoid even greater dangers in the future.

REFERENCES

1. Kevles DJ (1985) *In the Name of Eugenics*. Knopf, New York.
2. Adams MB (ed.) (1990) *The Wellborn Science. Eugenics in Germany, France, Brazil and Russia*. Oxford University Press, Oxford.
3. Mazumdar PMH (1992) *Eugenics, Human Genetics and Human Failings. The Eugenics Society, its Sources and its Critics in Britain*. Routledge, London.
4. Weindling P (1989) *Health, Race and German Politics Between National Unification and Nazism, 1870–1945*. Cambridge University Press, Cambridge.
5. Muller-Hill B (1988) *Murderous Science*. Oxford University Press, Oxford.
6. Harper PS (1991) *Huntington's Disease*. WB Saunders, London (2nd edition, 1996).
7. Jelliffe SE (1908) A contribution to the history of Huntington's chorea: a preliminary report. *Neurographs*, **1**: 111–124.
8. Punnett RC (1908) Mendelian inheritance in man. *Proc. Soc. Med.* **1**: 135–168.
9. Gusella JF, Wexler NS, Conneally PM *et al.* (1983) A polymorphic DNA marker genetically linked to Huntington's disease. *Nature*, **306**: 234–238.
10. Conneally PM, Haines J, Tanzi R *et al.* (1989) No evidence of linkage heterogeneity between Huntington disease (HD) and G8 (D4S10) *Genomics*, **5**: 304–308.
11. Meissen GJ, Myers RH, Mastromauro CA *et al.* (1988) Predictive testing for Huntington's Disease with use of a linked DNA marker. *N. Engl. J. Med.* **318**: 535–542.
12. Brandt J, Quaid K, Folstein SE *et al.* (1989) Presymptomatic diagnosis of delayed-onset disease with linked DNA markers: the experience in Huntington's disease. *J. Am. Med. Ass.* **261**: 3108–3114.
13. Meyer JE (1988) The fate of the mentally ill in Germany during the Third Reich. *Psychol. Med.* **18**: 575–581.
14. Davenport CB, Muncey DM (1916) Huntington's chorea in relation to heredity and insanity. *Am. H. Insanity*, **73**: 195–222.
15. Davenport CB (1911) *Heredity in Relation to Eugenics*. New York (reissued: Arno, 1972).
16. Harper PS (1991) The epidemiology of Huntington's disease. In: *Huntington's Disease*, pp. 251–280. WB Saunders, London.
17. Spillane J, Phillips R (1937) Huntington's chorea in South Wales. *Quart. J. Med.* **6**: 405–425.
18. Bell J (1934) Huntington's chorea. In: *Treasury of Human Inheritance* (eds R.A. Fisher, L.S. Penrose) Vol. 1, Part 1, pp. 1–64. Cambridge University Press, Cambridge.
19. Munchow R (1943) 950 Gutachten in Erbesundheitssache der Jahre 1934–1940. PhD thesis. University of Berlin, Berlin.
20. Hjardeng V (1991) Germany begins to lift the veil on HD. *Huntington Dis. Assoc. Newsletter*, **39**: 4–5.
21. Patzek B (1988) Sterilisations – Prozesse am Beispiel de Erbgesundheitgerichts Dusseldorf unter besonderl rucksichtigung der Chorea Huntington-Kranken. PhD thesis, University of Dusseldorf, Dusseldorf.
22. Panse F (1942) *Die Erbchorea, eine Klinische-genetische Studie*. Thieme, Leipzig.
23. Wendt GG, Drohm D (1972) *Die Huntingtonsche Chorea: eine Populations Genetische Studie*. Thieme, Stuttgart.
24. Schroeder-Kurth TM, Huebner J (1989) Ethics and medical genetics in the Federal Republic of Germany (FRG). In: *Ethics and Human Genetics* (eds D. Wertz, J. Fletcher), pp. 156–175. Springer, Berlin.

25. Erb-Sommer M, Muller LA (1987) *Stern*, **49**: 96–98.
26. Morris M, Tyler A, Lazarou L, Meredith L, Harper PS (1989) Problems in genetic prediction for Huntington's disease. *Lancet*, **2**: 601–603.
27. Morris M, Tyler A, Harper PS (1988) Adoption and genetic prediction for Huntington's disease. *Lancet*, **2**: 1069–1070.
28. Green J, McCreamer A (1989) *Counselling in HIV Infection and AIDS*. Blackwell, Oxford.
29. Nolan K (1991) Human immunodeficiency virus infection in women and pregnancy: ethical issues. *Obstet. Gynecol. Clin. North Am.* **17**: 651–658.

Chapter
18

CHINA'S GENETIC LAW

P.S. Harper

INTRODUCTION

On 1 June 1995, the People's Republic of China brought into force the first clearly eugenic law that the world has seen since modern genetics began to have impact on medical practice. Innocuously entitled the law on "Maternal and Infant Health Care", it contains, among other more general and uncontroversial proposals, clauses that are of profound significance for the application and perception of genetics far beyond the boundaries of China itself. Whether this law will lead to what would generally be considered the abuse of genetics remains to be seen, but the rulings would certainly legitimize, in the strict sense of the word, practice that would be unacceptable to the medical genetics community in most of the world.

Because of these wider implications, and because China itself contains one-third of the world's population, it is worth looking closely at this development; having myself been peripherally involved over a long period, and having found that many professionals in genetics are entirely unaware of the whole topic, I give here some background material that may help to put it into perspective.

First, it is relevant to quote (from the official Chinese translation) [1] some of the clauses in the law that specifically involve genetic disorders.

LAW OF THE PEOPLE'S REPUBLIC OF CHINA IN MATERNAL AND INFANT HEALTH CARE

Adopted at the 10th Meeting of the Standing Committee of the Eighth National People's Congress on 27 October 1994, promulgated by Order No. 33 of the President of the People's Republic of China on 27 October 1994, and effective as of 1 June 1995.

Article 8 The pre-marital physical check-up shall include the examination of the following diseases:

(i) genetic diseases of a serious nature;
(ii) target infectious diseases; and
(iii) relevant mental disease.

The medical and health institution shall issue a certificate of pre-marital medical check-up thereafter.

Article 10 Physicians shall, after performing the pre-marital physical check-up, explain and give medical advice to both the male and the female who have been diagnosed with certain genetic disease of a serious nature which is considered to be inappropriate for child-bearing from a medical point of view; the two may be married only if both sides agree to take long-term contraceptive measures or to take ligation operation for sterility. However, the marriage that is forbidden as stipulated by the provisions of the Marriage Law of the People's Republic of China is not included herein.

Article 16 If a physician detects or suspects that a married couple in their child-bearing age suffer from genetic disease of a serious nature, the physician shall give medical advice to the couple, and the couple in their child-bearing age shall take measures in accordance with the physician's medical advice.

Article 18 The Physician shall explain to the married couple and give them medical advice for a termination of pregnancy if one of the following cases is detected in the prenatal diagnosis:

(i) the fetus is suffering from genetic disease of a serious nature;
(ii) the fetus is with defect of a serious nature; and
(iii) continued pregnancy may threaten the life and safety of the pregnant woman or seriously impair her health due to the serious disease she suffers from.

Supplementary provisions

'Genetic diseases of a serious nature' refer to diseases that are caused by genetic factors congenitally, that may totally or partially deprive the victim of the ability to live independently, that are highly possible to recur in generations to come, and that are medically considered inappropriate for reproduction;

'Relevant mental diseases', refer to schizophrenia, manic-depressive psychosis and other mental diseases of a serious nature.

It could of course be argued, (and has been within China) that these proposals are simply the practical way of a country with relatively undeveloped services trying to ensure that prenatal diagnosis of genetic disorders and comparable measures are actually made available to its population; also that with a 'one child policy', such as already exists in China, it is important to ensure that the child born does not have avoidable handicap. It is certainly true that the law stipulates that decisions are to be made by appropriately trained people (article 26), while fetal sexing on non-medical grounds is specifically prohibited (article 32).

It is impossible though to deny the directive, even coercive tenor of the genetic clauses in the law, while its linkage with infectious diseases and mental illness makes it clear that genetic disorders are being considered primarily as a public health issue.

Why should China have produced a law of this type at a time when virtually all other countries have moved away from restrictive or eugenic legislation for genetic disorders? It is difficult for an outsider to be sure on this, but that broader political factors have been involved is clear from the following official commentary on the draft version of the law, produced a year before the final form [2].

HEALTH MINISTER PRESENTS EUGENICS LAW TO NPC STANDING COMMITTEE

(a) Xinhua new agency, Beijing in English 1114 Greenwich Mean Time (GMT), 20 Dec. 93.

Text of report

China is to use legal means to avoid new births of inferior quality and heighten the standards of the whole population. The measures include deferring the date of marriage, terminating pregnancies and sterilization, according to a draft law on eugenics and health protection which was presented to the current session of the Eighth National People's Congress (NPC) Standing Committee.

Explaining the law to participants at an NPC session that opened here today (Beijing, 20th December) Minister of Public Health, Chen Minzhang, said that the measures will help prevent infections and hereditary diseases and protect the health of mothers and children.

Under the draft law, those having such ailments as hepatitis, venereal disease or mental illness, which can be passed on through birth, will be banned from marrying while carrying the disease. Pregnant women who have been diagnosed as having certain infectious diseases or an abnormal foetus will be advised to halt the pregnancy. Couples in the category should have themselves sterilized, the draft says.

China is in urgent need of adopting such a law to put a stop to the prevalence of abnormal births. Minister Chen explained statistics show that China now has more than 10 million disabled persons who could have been prevented through better controls.

The draft also stipulates that organizations that are engaged in pre-marital checks, eugenics, pre-birth diagnosis or sterilizations should be approved by the authorities at the county level and above. Chen said, "Personnel involved in this area should be subjected to strict training".

The Minister of Public Health called on medical authorities at various levels to establish a comprehensive network for the implementation of the law.

The draft does not state whether China will adopt euthanasia to eliminate congenitally abnormal children, saying that the international community has not come to a conclusion on that issue. The draft also does not touch on the issues of artificial fertilization or test-tube babies because the effects of these techniques have caused some disputes and because it's too early to put any limitations into law.

Minister Chen said the government should strengthen its control and supervision of these techniques. He called for the Ministry of Public Health to work out tentative management measures of control.

(b) Xinhua new agency domestic service, Beijing, in Chinese 0731 GMT, 20 Dec. 93.

Excerpts from report

Premier Li Peng submitted five bills to the fifth meeting of the English National People's Congress (NPC) Standing Committee, which opened today [20th December] for deliberation by the Standing Committee . . . [see accompanying report in this section for note of the five bills submitted].

Explaining the draft law on eugenics and health protection, Minister of Public Health Chen Minzhang said:

Since the founding of the NPC, we have done a great deal in publicizing to the people and educating them on eugenics and health protection and we have pioneered a number of measures for better quality births. However, due to the lack of legal protection for eugenics and health protection work, as well as economic backwardness and the lingering influence of outdated thinking, abnormal births are still rather prevalent in China. According to statistics, of the five categories of handicapped people in the country, more than 10 million are disabled at birth, constituting 9.9 per 1000 of the population; among them, 4.17 million are children under 14. There are 10.17 million mentally retarded patients nationwide, about 10 per 1000 of the population of children under 14, about four million have low IQs, accounting for 10.7 per 1000 of the children of that age group. Each year about 300 000–460 000 children are born disabled because of hereditary diseases (only those with visible defects), or 13 to 20 per 1000 of new births. The perinatal mortality rate is 26.17 per 1000; of this, one-fifth of the deaths are caused by hereditary diseases.

Chen Minzhang also said:

Births of inferior quality are especially serious among the old revolutionary base, ethnic minorities, the frontier and economically poor areas. In some villages, no competent people can be found to work as accountants or cadres or can be recruited by the army because of long-term isolation, backward production, consanguineous marriages and excessive births. The cost of nursing, caring and providing medical treatment and other services for the 400 000 births with hereditary handicaps each year is enormous. Because China will experience its third population birth peak during the Eighth Five-Year Plan period, births of a large number of people with mental retardation or hereditary diseases will inevitably impose additional burdens on the state and bring misfortune to tens of millions of families. In addition, a sample survey shows that most of the couples who apply to have a second child do so because their first child is born handicapped or with hereditary diseases. If this situation continues for a long time, the quality of the Chinese population is bound to deteriorate markedly.

Chen Minzhang went on to say:

The state of inferior-quality births has aroused grave concern in the whole society and their latent effects have alarmed and worried people in various circles. Currently, the broad masses of the people demand that a eugenics law be enacted and effective measures be taken to reduce inferior-quality births as quickly as possible. The previous sessions of the NPC and the Chinese People's Political Consultative Conference National Committee made motions, proposals and suggestions for expediting legislation on eugenics. Therefore, it is necessary to formulate as soon as possible a law on eugenics and health protection and to ensure better-quality births and to control and reduce inferior-quality births . . . .

These quotes clearly show the eugenic and even racial thinking behind the law, and that the issues are being seen primarily in political and economic terms. It seems unlikely that such comments were drafted in consultation with professionals in medical genetics; indeed one eminent medical geneticist in China to whom I wrote at this time had no knowledge whatever of the proposals and was greatly concerned to learn of them. Widespread international criticism of the draft proposals, notably in *Nature* [3], may possibly have contributed to the removal of all references to eugenics in the final law, even though its substance would seem to have changed little.

The story of China's genetics law goes back much further than 1993, however; since the background is known to very few outside China, it is worth putting on the record the facts as told to me during a visit to the principal Chinese medical genetics centres in 1986. During the course of this, I was both surprised and disturbed to be given a list of genetic disorders that were to be included in a proposed genetics law prohibiting marriage, and to be asked whether I agreed with this. Further informal discussion produced the following history, which I wrote down in detail the same night, for fear that accuracy might be lost if an immediate record were not to be made. If it is indeed erroneous, the responsibility is mine alone.

Medical Genetics in China was essentially non-existent until 1976, initially because of the influence of the Soviet Union and Lysenko, subsequently because of the cultural revolution, in which most professionals were humiliated and lost their jobs. When work in the field became feasible, there was initial disapproval of eugenics because of its association with Nazi Germany, but this changed rapidly, with a genetics law being first proposed around 1979. In 1980, a law prohibiting first cousin marriages was passed, following which more extensive laws were planned. At this point, the Government asked medical geneticists for their views; apparently there were five special meetings on the topic up to 1986, with the substance of the law containing three types of prohibition, much as in its final form; these were banning of marriage, banning of any reproduction but not marriage, and banning of a second child in a situation of high recurrence risk.

The list of disorders involved was apparently passed round different specialties, with each adding more disorders, so that the eventual list was very long.

I was told that there had been much debate and disagreement about the desirability and scope of the laws, which had been drawn up by non-medical geneticists. Medical geneticists mostly wished them to be as limited as possible.

At the time of my 1986 visit, the issue was clearly sensitive and unresolved. I was told that an opinion from abroad would be helpful, especially since my book on genetic counselling had been translated into Chinese [4]. I therefore wrote the following 'open letter' to the Chinese Society for Medical Genetics, via a distinguished Chinese medical geneticist, now living abroad. Since the issues seem as valid today as when it was written 10 years ago, I give it in full, even though it may appear rather formal in style. (I should also acknowledge that my own views on the 'preventive' aspects of genetics have also become more cautious over the subsequent 10 years.)

It was a very great pleasure for me to meet you and to see the excellent work that you are doing. During our discussions concerning developments in Medical Genetics and the possible eugenics laws, you asked me to write down my thoughts on this subject, so I am doing this in this letter. Of course these are my personal views, though I think that many of my colleagues in Europe and America would largely share them. Also, after visiting China only briefly, and having only such small knowledge of your country and its social system, it seems quite inappropriate of me to suggest what is best to be done in China. However, these are my thoughts, which you are free to use as you wish, and to share with your colleagues. If they are of some help I shall be greatly pleased and honoured.

Before discussing the question of eugenics laws, I should like to mention the issue that is most striking to a visitor from the West, and of course is well recognised by yourselves; that is the very serious population problem that you have. I was most impressed by the way China is facing up to the great difficulties produced by this problem, and by the work of the Family Planning Bureau and other bodies in trying to find solutions. Although the idea of a one-child family is unfamiliar and not easy to accept for someone from the West, I fully recognise that this measure is a necessity at the present time for China, and that if it were not to be enforced then even greater difficulties would occur in future. This is a situation quite different to what we have in the West, and it clearly requires a special solution.

Turning to genetic disorders, however, it would seem from what I have seen, and from discussions with geneticists and others, that the type and the scale of this problem is not greatly different to that seen in many western countries. Clearly there are some disorders which are more common and others more rare, but overall, the situation seems to me to be very similar to that which faced us some 15–20 years ago when the subject of medical genetics began to develop. As infectious diseases are controlled, genetic disorders become more conspicuous; this is again a situation you are seeing now and that we saw previously.

Because of this broad similarity, I would question whether China requires an approach to its genetic problems (in contrast to its population problems) that is different from that which has proved and is still proving remarkably effective in most other countries. I think it is worth looking at the major genetic problems in turn to see how they have been approached in countries like my own, and whether such approaches are likely to be successful in China.

The first point to be emphasised is how rapid is the pace of advance in our understanding and prevention of genetic diseases. Already many diseases are preventable where even a few years ago this would have seemed unlikely; this rapid advance will certainly continue, and I have seen for myself how Chinese geneticists are themselves using these techniques.

Considering chromosome disorders, some of the most important causes of mental retardation, these are already preventable where a risk is known, by means of prenatal diagnosis, especially now that chorion villus biopsy in early pregnancy is available. It is still difficult to prevent the first case in a family, but advances can be expected here also. Turning to single gene disorders, the thalassaemias are the major problem in many parts of China, and these are already preventable by a combination of prenatal diagnosis and heterozygote screening. Many countries have already set up systematic population screening programmes to detect high-risk couples before a first affected child is born, and in those countries (including less developed countries) the incidence has fallen rapidly. This would seem an area suitable for development in those parts of China with a high frequency of these disorders.

Many other serious single gene diseases are now detectable prenatally, either by biochemical measurements, or by the new DNA markers. Every month this becomes possible for still more disorders, so that most of the serious single gene disorders should be able to be prevented from recurring in a family within a very few years. Although these new techniques are expensive at present, the cost is falling with development of simpler methods for processing large numbers of samples.

Polygenic disorders are a difficult problem because of their high frequency, and are of course only partly genetic. Here also, some of the most important can be prevented, such as neural tube defects, while others such as congenital heart disease should be detectable with improved ultrasound techniques, such as you are already introducing.

For all these groups of genetic disorders, genetic counselling plays an important role, in making both ordinary people and medical personnel aware of the possible risks of an affected child and what can be done for prevention. I think that with your one-child family policy, genetic counselling is especially important and I am sure that education and information will do much to encourage people to seek advice. In my own country there has been a rapid increase in demand for genetic counselling and in public awareness of its importance, and I am sure that you will find the same. I was pleased to learn that courses in genetic counselling for your doctors are in progress, and expect that you will also need more specialists in this field.

Taking the situation overall, I feel that the prospects for a considerable reduction in genetic disorders in China are good, using the same developments that have been effective in the West. We have found that the results of these programmes have been cost-effective in economic terms, as well as having beneficial social effects and helping the families concerned.

This leaves the question of eugenic or genetic laws, which is a most difficult issue, and one which I know has been the subject of much careful discussion in your country. I am not a legal expert, but I see laws as being of two main types – those which promote and encourage certain activities thought to be helpful to society, and those which forbid or restrict certain activities considered to be harmful.

I can see great benefit coming from a genetics law in China of the first type, which could set standards for establishing satisfactory genetics services throughout the country, could encourage education and public information, and provide support for development in this field. It seems to me that at present, the genetics services that exist are rather fragmented and uneven in distribution, and that they often depend on the initiative of a small number of able people without extensive support. Of course, economic considerations may limit the extent of developments and priorities would need to be set, but a law would certainly help in this direction.

Concerning the second category of law, that would restrict the marriage or childbearing of individuals with particular genetic disorders, I have the gravest reservations about this, for a number of reasons. First I cannot see how it would be effective in preventing genetic disorders by comparison with the approaches I have described above, which have all been so successful in the West. If a combination of preventive tests and genetic counselling have worked well in other countries with rather similar genetic problems to China, in the absence of restrictive laws, then I think they will also work well in China without such laws. Indeed this already seems to be happening where the facilities are available.

Finally, I am concerned that the introduction of restrictive laws could have serious implications for relationships between workers in Medical Genetics in China and colleagues in other countries. Almost all those countries that in the past had eugenics laws have now abandoned them, partly because they were ineffective and unworkable, partly because they were felt to be inappropriate and did not have popular support. I believe that only in Japan and possibly South Korea do such laws persist; both are rather rigid and traditional societies, and medical genetics is not well developed in either to my knowledge. Because of the strongly held feelings of many western workers on this subject, there is a real possibility that the introduction of restrictive laws might hinder the development of contacts with medical geneticists in Europe and America and could impede the free flow of new techniques and materials. I should personally be extremely sad if this were to result, having experienced myself how fruitful and mutually beneficial such contact is.

Thus my strong feeling is that restrictive eugenic or genetics laws would be both ineffective and harmful, whereas a law designed to promote and enhance the development of genetics services could, if carefully planned, be of great positive value in the prevention of genetic disorders. From the impressive work that I have seen myself during my visit, the foundations of these genetics services have already been laid, and their further development offers the prospect of considerable and increasing success.

Secondly, I think there would be very serious difficulties in defining suitable categories of genetic disorders for any laws. Genetic diagnoses are often difficult, and there could easily be confusion and mistakes. Agreement as to which diseases are serious and which less serious is also difficult, as is agreement as to what is a high genetic risk. Also, the rapid pace of genetic developments means that any list of diseases will soon become outdated and inappropriate; it takes time to alter a law once it is made. To give an example, bilateral retinoblastoma is a serious inherited eye disorder with a high risk to children, but this year prenatal diagnosis has become possible with identification of the genetic locus. To have placed this disease on any list for prohibiting marriage or childbearing would be wrong now it is becoming preventable. Many other serious diseases will soon become preventable in a similar way.

Thirdly, I would fear that restrictive laws would have a serious adverse effect on the programmes of education and public involvement that are essential for the prevention of genetic diseases. In my own work on preventing Huntington's chorea and in other programmes, it is absolutely essential to have the support and involvement of the public if success is to be achieved. A fear that they might be refused permission to marry or have a child may result in people concealing a disorder or a positive family history; this could make it very difficult to carry out accurate studies of genetic disorders and could hinder efforts in prevention that are currently in progress.

I must apologise for the length of this letter, and for raising issues which I am sure you have already discussed fully; I greatly look forward to seeing the progress of your country in this field, and know that there is much goodwill around the world for the efforts that are being made.

I have no means of knowing what, if any effect this letter may have had, nor whether the background to the genetics law given here is strictly correct since the facts came from a single (though very well informed) person. In 1992 I was asked to write an editorial on social and ethical issues for the *Chinese Journal of Medical Genetics* [5], but I heard nothing more on the topic of laws until the 1993 draft proposals, already quoted, appeared. It is clear that the final passage of the genetics law was the culmination of 15–20 years of intense debate, almost entirely confined to China, and that the issue is probably not yet closed.

To me, this episode shows that genetics as applied to medicine is always likely to be politically a sensitive, and at times dangerous area, especially when it appears to offer politicians and health planners solutions to intractable and expensive health problems. China is far from unique in this respect.

What has happened in the 2 years since the law was passed? There has been strong criticism in UK-based journals (*Nature* [6], the *Lancet* [7], *Journal of Medical Genetics* [8]) and from the UK Clinical Genetics Society [9], but none from the American Society of Human Genetics or its journal. At least one international genetics meeting in China (on hereditary ataxias) has been cancelled following protest from the lay society concerned, and it has been questioned whether the next International Genetics Congress, due to be held in Beijing in 1998, should proceed. Following a ballot of its members, the UK Genetical Society has withdrawn from the International Genetics Federation on this issue [10]. Understandably, people vary in their views as to whether closer contacts or boycott is the policy most likely to have an effect.

My personal view is that the implications of China's genetic law are truly international, and that it would severely damage the standing and perception of medical genetics world-wide unless genetics professionals are prepared to make their disapproval and opposition clear. I do not think it is adequate to excuse the law as simply another facet of an authoritarian regime, or to oppose breaking off of collaborations, on the ground that this might hurt those inside China who are trying to help the specialty to evolve. The law, and the preceding lists of 'prohibited disorders' have an uncanny and disturbing resemblance to the infamous Nazi 'Law of 1933', a law also largely drafted by geneticists and which was the foundation of greater damage to patients and families with genetic disorders than any other event since the beginning of genetics. Unless we as professionals in the field make our views strongly felt, something that we can do with more impunity than our Chinese colleagues, the public world-wide will rightly ask "why were they silent?"

REFERENCES

1. Legislative Affairs Commission of the Standing Committee of the National People's Congress of the People's Republic of China (transl.) Law of the People's Republic of China on Maternal and Infant Health Care, pp 1–13.
2. Health Minister presents eugenics law to NPC Standing Committee. Xinhua News Agency, Beijing, 22 December 1993.
3. Mao X (1994) China's misconception of eugenics. *Nature*, **367**: 1–2.
4. Harper PS (1986) *Practical Genetic Counselling* (Chinese translation). Shanghai.
5. Harper PS (1993) Social and ethical aspects of molecular genetics. Problems in the West and their relevance to China. *Chin. J. Med. Genet.* **10**: 193–197.
6. Editorial (1995) When is prenatal diagnosis 'eugenics'? *Nature*, **378**: (7 December).
7. Editorial (1995) Western eyes on China's eugenics law. *Lancet*, **346**: 131. (See also succeeding correspondence.)
8. Bobrow M (1995) Chinese law remains eugenic. *J. Med. Genet.* **32**: 409.
9. Burn J, Jacobs PA, Berry AC, Modell CB, Patch C, Sykes B (1995) Concern at China's new genetic law. *The Times*, June 5.
10. Genetical Society (1996) International Congress of Genetics in China: what should the Genetical Society do? *Genet. Soc. Bull.* **30**: 9–11.

FURTHER READING

In writing this book, we have not been seeking to demonstrate originality. In part, we have wanted to identify issues that seem to us, as clinicians, to be of particular importance. In addition, we hope to give a clinical perspective on issues that have already been raised by others. It is therefore appropriate for us to list some of the contributions made by others to the discussions that are continued within this book. We acknowledge some of these contributions as appropriate in specific chapters, but have decided to list here some of the books and papers that can serve as a general guide to the issues raised throughout this volume.

BOOKS AND REPORTS

Abramsky L, Chapple J (eds) (1994) *Prenatal Diagnosis: the Human Side*. Chapman & Hall, London.

Andrews LB, Fullarton JE, Holtzman NA, Motulsky AG (eds) (1994) *Assessing Genetic Risks: Implications for Health Policy*. National Academy Press, Washington, DC.

Annas GJ, Elias S (1992) *Gene Mapping: Using Law and Ethics as Guides*. Oxford University Press, Oxford.

Bosk CL (1992) *All God's Mistakes: Genetic Counselling in a Paediatric Hospital*. University of Chicago Press, Chicago.

Clarke A (ed.) (1994) *Genetic Counselling: Practice and Principles*. Routledge, London.

Clarke A (ed.) (1997) *Culture, Kinship and Genes*. Macmillan, London.

Duster T (1990) *Backdoor to Eugenics*. Routledge, Chapman and Hall, London.

Harper PS (1993) *Practical Genetic Counselling*, 4th Edn. Butterworth-Heineman, London.

Kevles DJ (1986) *In the Name of Eugenics*. Knopf, New York.

Kitcher P (1996) *The Lives to Come: the Genetic Revolution and Human Possibilities*. Penguin Books, London.

Lewontin RC (1991) *The Doctrine of DNA: Biology as Ideology*. Penguin Books, London.

Marteau T, Richards M (eds) (1996) *The Troubled Helix: Social and Psychological Implications of the New Genetics*. Cambridge University Press, Cambridge.
Nelkin D, Lindee MS (1995) *The DNA Mystique: the Gene as a Cultural Icon*. WH Freeman and Co., New York.
Nuffield Council on Bioethics (1993) *Genetic Screening – Ethical Issues*. Nuffield Council on Bioethics, London.
Rothman BK (1988) *The Tentative Pregnancy*. Pandora, London.
Royal College of Physicians Committee on Ethical Issues in Medicine and Clinical Genetics (1991) *Ethical Issues in Clinical Genetics*. Royal College of Physicians of London, London.
Wilkie T (1993) *Perilous Knowledge: the Human Genome Project and its Implications*. Faber and Faber, London.

PAPERS

Chadwick R, Levitt M (1996) EUROSCREEN: Ethical and philosophical issues of genetic screening in Europe. *J. Roy. Coll. Phys. London*, **30**: 67–69.
Green JM, Richards MPM (eds) (1993) Psychological aspects of fetal screening and the new genetics. *J. Reprod. Infant Psychol.* **11(1)**: 1–62 (special issue).
Holtzman NA (1992) The diffusion of new genetic tests for predicting disease. *FASEB J.* **6**: 2806–2812.
Kessler S (1989) Psychological aspects of genetic counselling: VI. A crucial review of the literature dealing with education and reproduction. *Am. J. Med. Genet.* **34**: 340–353.
Lippman A (1992) Led (astray) by genetic maps: the cartography of the human genome and health care. *Soc. Sci. Med.* **35**: 1469–1476.
Macintyre S (1995) The public understanding of science or the scientific understanding of the public? A review of the social context of the 'new genetics'. *Publ. Underst. Sci.* **4**: 223–232.
McLean SAM (1994) Mapping the human genome – friend or foe? *Soc. Sci. Med.* **39**: 1221–1227.
Richards MPM (1993) The new genetics: some issues for social scientists. *Soc. Hlth Illness,* **15**: 567–587.
Wertz DC (1992) Ethical and legal implications of the new genetics: issues for discussion. *Soc. Sci. Med.* **35**: 495–505.

INDEX